BIRDS AT REST

Birds at Rest

THE BEHAVIOR AND ECOLOGY
OF AVIAN SLEEP

ROGER F. PASQUIER

ILLUSTRATED BY
MARGARET LA FARGE

PRINCETON UNIVERSITY PRESS
PRINCETON & OXFORD

Published by Princeton University Press
41 William Street, Princeton, New Jersey 08540
99 Banbury Road, Oxford OX2 6JX

press.princeton.edu

All Rights Reserved

ISBN 9780691259963
ISBN (e-book) 9780691260075

British Library Cataloging-in-Publication Data is available

Editorial: Robert Kirk and Megan Mendonça
Production Editorial: Mark Bellis
Jacket Design: Wanda España
Production: Steve Sears
Publicity: Matthew Taylor and Caitlyn Robson-Iszatt
Copyeditor: Leah Caldwell

Jacket Credit: Greater Flamingo, Helge Sørensen / AGAMI

This book has been composed in Arno

Printed in the United States of America

10 9 8 7 6 5 4 3 2 1

For Hugo and Eloise van Lint
the next generation of naturalists
in our family

CONTENTS

Is it common for birders to take an interest in how and where birds sleep? As a child in New York City, I used to enjoy watching the pigeons settle down for the night on the ornamented ledges of a building across from me on East 73rd Street. When old enough to come home from school by myself, I sometimes detoured slightly during the darkest afternoons of early winter to see the flocks of starlings land in swirls on their roost at the Metropolitan Museum of Art. In Paris as a teenager, my detour was to the Louvre, where during the summer dusk I enjoyed the contrast of the House Martins returning to their mud nests under the cornices while the Common Swifts soared higher and higher until out of sight. Back in East Hampton, Long Island, at the end of those summers, I went to see the flocks of Tree Swallows circling and then settling into the reeds at the edge of Hook Pond; in September, this conveniently happened just before I had to be in for dinner.

Over the years as an adult in New York, I discovered that there were many roosting spectacles in the neighborhood and for much of the year: from June through March, starlings and Common Grackles heading down Fifth Avenue to the trees around the Pulitzer Fountain at Grand Army Plaza; from August into mid-October, hundreds of migrating Chimney Swifts whirling and then dropping into chimneys atop buildings on Fifth and Park avenues, just about when the street lights go on; on the East River Esplanade, in late afternoon, Common Crows singly or in small groups flying from the Bronx to gather on the roof of a warehouse and then taking off in a flock to their actual roost somewhere beyond sight in Queens. Also, leafy trees noisy all night with House Sparrows, or the grove of elms in Central Park where, long before dark in late summer, American Robins come, some from across the Hudson River.

If all these evening diversions exist in New York City, the varied spectacles of birds coming to roost must be present everywhere. Everyone
who enjoys this will have personal favorites. Mine, perhaps, is the many
evenings I have spent on a platform at the top of a ceiba tree towering over
the rainforest in southeastern Peru as the brief tropical dusk falls. The last
White-throated Toucan gives its final yelps, the plaintive whistles of the
Undulated Tinamous fade out, the Starred Wood-Quails call their rollick-
ing crepuscular duet, and the first Ocellated Poorwill may purr. It is then
that twos and threes of Blue-and-yellow Macaws leisurely fly high over
the canopy, their long central tail feathers characteristically undulating.
The locals say the macaws are on their way to Paititi, the legendary lost
Inca city of gold.

Beyond these spectacles, sleep and roosting habits give one much to
think about. My interest was piqued when, at age thirteen, I read David
Lack's account of the Common Swifts in Oxford, with his speculations
on what non-nesters do at night, citing the early evidence that they
spend it aloft. Decades later, when writing *Birds in Winter*, I learned the
various ways small birds in northern forests survive the long and extremely cold nights and how shorebirds adjust their resting schedules
to the shifts of the tides. One day, working on that book in the library
of the Department of Ornithology in the American Museum of Natural
History, I was reading about Rock Ptarmigans on Svalbard, a Norwegian archipelago at 74–81°N, in the Greenland Sea. The Rock Ptarmigans live through ten weeks lacking almost all light, and they spend
most of each day burrowed in snow. That got me wondering if one could
write an entire book about how birds sleep. Putting down my pencil,
I looked around the shelves for anything on the subject. I found Alexander
Skutch's 1989 *Birds Asleep*. Fascinating, full of his characteristic detailed
firsthand observations, valuable insights, and useful references to the
literature, mostly anecdotal. But nothing on the evolutionary origins of
sleep, its physiology and functions, the energetics of commuting to a
roost, the thermal benefits of different sites, and other topics that today
can be discussed in detail. I knew, however, that behind the bindings of
the rows and rows of books and journals on the library's long shelves
must be much more, and still more in the museum's main library, plus

all that now comes so easily on Google Scholar—if one knows what to search for. That inspired my next treasure hunt, launched when *Birds in Winter* was finished.

This book is the result. The more I read and wrote, the more I saw that understanding sleep and its associated behaviors, like daytime rest and the specific habitat features birds seek for their inactive hours, is essential to appreciating any species' ecology, behavior, and conservation needs, its daily cycle and life history. Inevitably, on the way, I found that avian sleep and its associated roosting habits are much more complex—may I say sophisticated?—than I had imagined. I hadn't known that sleep is universal in the animal kingdom, with even jellyfish and other creatures lacking a brain regularly taking a break every day. Or that among birds, some can forgo sleep for extensive periods when other activities have priority. Or that birds sleep in a long series of bouts that each run only several seconds or a few minutes. And that they routinely sleep some of that time with only half their brain at rest while one eye and the other half are alert. Or that some birds even do this on the wing. While roosting singly in a sheltered spot would have been mastered by birds' dinosaur ancestors (a fossilized non-ancestor dinosaur has been found in the same sleep posture used by most birds), and roosting in flocks would come naturally to birds that nest in colonies, I wondered how the communal roosting habit evolved among sedentary and migratory birds that spend the breeding season unsociably on territories. How did some evolve the practice of swirling in dense flocks and then plunging en masse to their roost to evade aerial predators? What about those that nightly drop their body temperature several degrees to conserve energy, or go even further, entering torpor, waking slowly in the morning? Others, meanwhile, never land for months but still function perfectly. And how did some birds learn to go to caves during the hottest hours of the day, or to crevices in glaciers during the coldest nights?

These years were a good time to ask such questions, as so much new technology has become available. We can now measure a bird's body temperature unobtrusively as it sleeps, record its brain waves as they shift between slow-wave sleep and rapid eye movement sleep, and follow individual birds as they move across the planet wearing tiny devices that track their heart rate, note if they are active or still, on land or in water. Meanwhile, laboratory experiments set out to learn whether pigeons have to sleep longer after kept awake past their bedtime (watching David Attenborough's *The Life of Birds*); if tits can detect and would then decide to avoid a nest box impregnated with various signs of a predator; and how different types of artificial light now being installed in urban and suburban areas affect birds' metabolism, reproductive cycle, and health. With fascinating and unexpected results coming both from recent field observations and labs, I found it hard to stop checking the many ornithological, behavioral, ecological, evolutionary, and physiological journals where new findings are reported. I drew the line at the end of 2023, knowing that future journal issues would have information I wished I could have included.

This book is therefore a mix of modern science—if we think of science as the testing of hypotheses—and more old-fashioned natural history. That is, the direct observation and description of birds going about their lives—the kind of engagement that anyone interested in birds today can enjoy with no equipment at all. Each chapter likely has a different mix of the two. I begin with the evolution and functions of sleep and its several components in birds. Chapter 2 describes the additional varieties of rest that birds use, from midday breaks during the hottest hours to hypothermia and torpor, both deployed to get through the coldest nights. With the mechanics of sleep covered, the next two chapters focus on where birds roost to give themselves safety from predators and shelter from the elements. This takes two chapters because, while most birds may seek the same features, others have evolved more specialized behavior, such as building dedicated roost structures in trees, into the ground, and through snow, among other places, or by abandoning the terrestrial world entirely and spending the night in the air.

Sleep and roosting also have their social aspects. Chapters 5 and 6 describe how birds use their roosting time to find a mate, maintain a pair bond, protect their family, or learn—among other things—where to get breakfast. Birds are more regular in their sleep habits than we are. Chapter 7 discusses roosting times and how they are affected by light levels, day length, phases of the moon, temperature and extreme weather, and by each bird's foraging success. Chapter 8 describes the more elaborate behavior of birds that commute to massive communal roosts. Despite birds' efforts to find a safe place to sleep, they are inevitably preyed upon by a wide range of animals, including insects, snakes, other birds, and mammals. This, and how birds compete for their roost sites with other species, are covered in chapter 9. Our own species has also impacted bird sleep: artificial light and noise; roost disturbance in the forms of traffic, hunting, and destruction; the invasive species that now kill birds when they are least able to respond; plus, inevitably, climate change. These are all be acknowledged in chapter 10.

In addition to discovering the exciting revelations modern science has made possible about sleep and roosting habits, in the course of writing this book I especially enjoyed finding observations by immortals in the pantheon of ornithology and natural history. Some, like Alexander Wilson, John James Audubon, and William Brewster, saw phenomena we no longer can. Others, including Gilbert White, Alfred Russel Wallace, and Niko Tinbergen, were pioneers speculating on aspects of bird sleep. I even came across valuable observations by Cotton Mather and Theodore Roosevelt.

I also found that I had watched birds roosting in some of the same places as my grander predecessors. From Joseph Hickey's still valuable *A Guide to Bird Watching* (1943), I learned that both he and, a few years later, Roger Tory Peterson, had tried to count the starlings at the Metropolitan Museum. On December 5, 1938, Peterson estimated thirty-eight thousand. A friend reminded me that Chaucer, in his prologue to *The Canterbury Tales* (1386), wrote "And smale foweles maken melodye / That slepen al the nyght with open ye." Did he know about unihemispheric sleep, and how would that have been written in his century's English? My hope is that today's readers, as they learn about and observe birds resting, sleeping, or preparing to roost, will enjoy the sense that they are participating in a long tradition, and that they are witnessing the result of millions of years of evolution. There is still so much to discover.

ACKNOWLEDGMENTS

My first debt is to the scientists and naturalists who have spent serious time in the field and in labs, observing and experimenting to understand the nature of avian sleep and the habits associated with it. Their original research and even their chance observations have provided a rich lode from which I have sought to find patterns and exceptions, to pick the most vivid examples and most impressive advances.

As with the writing of three previous books, I am grateful for the hospitality and resources of the Department of Ornithology at the American Museum of Natural History. In addition to the riches of the department library, I have always relished the privilege of looking at specimens of birds I didn't already know, whenever I encountered them in my reading. In the department, George Barrowclough shared useful papers, Paul Sweet introduced me to the Biodiversity Heritage Library, Tom Trombone solved Internet problems, and Augie Kramer and Vicens Vilà-Coury explained the landscape of the Cornell University campus. At the museum's main library, Mai Reitmeyer and Annette Springer graciously retrieved books and journals from the stacks during the Covid and construction years, when the rest of us did not have access.

Alan Poole, former editor of *The Birds of North America*, kindly secured for me from the Cornell libraries a few journal papers that neither the American Museum nor Google Scholar seemed to have. Nigel Collar, my former colleague at BirdLife International, found information on Amur Falcon stopover sites in India and on cat and rodent eradication projects on various oceanic islands. Danielle Whittaker of the COLDEX Center for Oldest Ice Exploration at Oregon State University directed me to her important papers on the social functions of avian olfaction. Richard Prum of Yale University introduced me to *Mei long*, the non-avian dinosaur that some 128–139 million years ago was asleep when volcanic ash covered it and preserved it in the same position many birds use today. Grant Sizemore of the American Bird Conservancy gave me the most recent figures (appalling) on the number of birds killed annually by cats in the United States and several other countries. Steven Brown and Brian Harrington of the Manomet Observatory updated me on the current state of shorebird protection on Plymouth Beach, Massachusetts. Whit Stillman, *cinéaste*, scanned the French literature. Susan Ferriere reintroduced me to Chaucer. Niels Rattenborg of the Avian Sleep Group at the Max Planck Institute for Animal Intelligence, the author of more papers than any other scientist whose work I drew upon, generously read the entire manuscript. An anonymous reviewer also made valuable comments. I, however, remain responsible for any errors of facts or interpretation.

Margaret La Farge's pencil drawings make vivid some of the birds and phenomena described in the text. I wish there could have been more!

At the Princeton University Press, Robert Kirk, editor; Mark Bellis, associate managing editor; Leah Caldwell, copyeditor; and David Luljak, indexer, all applied their skills to bringing this book to material existence. I benefited from their expertise and am grateful for what I learned from them.

BIRDS AT REST

1

Why Birds Sleep

Sleep is an essential part of life for jellyfish, snails, octopuses, insects, fishes, and reptiles through to birds and mammals. Indeed, animals could not survive without sleep, even as the portion of each day and of their lifetime in this state differs widely among classes and species. Timing, duration, and intensity of sleep vary in response to each animal's habitat, daily activity, reproductive state, place on a food chain, and other ecological and behavioral factors.

Since animals are often at greatest risk when asleep, sleep must provide benefits that compensate for the danger. Animals that are frequent prey for others have evolved different sleep patterns than predators, just as those sleeping in exposed places sleep differently from those in secure ones. Similarly, animals living at latitudes with extended daylight or long nights use those hours differently. Migratory animals, such as many birds, alter their sleep regime when traveling and at their destinations.

Sleep's key visible feature is periods of immobility, usually spent in a characteristic posture, during which responses to external stimuli are lower. Among animals in which brain activity has been measured, this also changes. In birds and mammals, with their complex ecologies and behaviors, sleep's aspects and functions increase. The brains and bodies of birds require substantial energy to operate in environments that challenge their ability to find food, to reproduce, to avoid dangers, and, for some, to travel long distances. Rest, when they can conserve and restore energy, therefore has more features than in animals with fewer daily demands.

Functions of Sleep

Sleep's functions and benefits reflect both ecological adaptations and neurological features. For most animals, there are portions of the twenty-four-hour cycle in which light levels make continuation of feeding, other maintenance activities, or reproduction more difficult. During the hours that wakefulness is unproductive and may even be dangerous, inactivity, if not actual sleep, is the most efficient use of this time. For warm-blooded animals—birds and mammals—sleep can also reduce the substantial costs of maintaining endothermy, a higher body temperature than the surrounding atmosphere. In sleep, many endothermic animals lower their body temperature, a significant energy saver that reduces their need for the food that fuels their high metabolic rate. The greater complexity of sleep mechanisms in endotherms may have evolved with warm-bloodedness itself, as a way to offset endothermy's high costs (Berger and Phillips 1995).

Sleep, when the brain is not substantially engaged in directing activity, is also the opportunity for it to restore itself to maintain high waking neurophysiological performance. During sleep the brain strengthens or weakens synaptic connections, effectively creating space for consolidation of learning and for organization of memory. Sleep may also cleanse the brain of waste and potentially damaging byproducts generated by neuronal activity during wakefulness (Rattenborg and Gonzalez-Martinez 2014). When birds are active, the major activity of their brain is the processing of sensory information, primarily visual. During sleep, when sensory experience and processing are substantially reduced, the brain can refresh memory circuits without conflicting demands.

In the course of animal evolution, the processing of sensory information in brains of ever greater complexity, such as those of birds, created more need for refreshing memory circuits. Sleep became all the more necessary (Kavanau 1998). Laboratory experiments have shown that sleep is involved in, among other things, consolidating imprinting memories for chicks, song learning in finches, auditory discrimination in starlings, and courtship in pigeons. At the same time, some memories may be shed during sleep, preventing saturation in the brain's hippocampus, where memories are stored (Rattenborg et al. 2010; Ungurean et al. 2021). Memory and learning are affected by quality as well as quantity of sleep. For the chicks in memory experiments, nine hours of sleep after imprinting training generated a lasting memory. Six hours of undisturbed sleep achieved almost the same result, but chicks given six hours of disturbed sleep immediately after training did not form any memory of the imprinting stimulus (Vorster and Born 2015).

In oscines ("songbirds") and suboscines, the muscles controlling the syrinx have bursts of activity during sleep that resemble fragments of the activity during song production; this takes place without respiration, so produces no sound. It may, however, contribute to creating useful muscle memory for subsequent song production (Döppler et al. 2021). In the brain as well, patterns of neural activity during sleep can match what the brain does during certain waking states, thereby reinforcing the motor patterns needed to perform the activity. Zebra Finches (*Taeniopygia guttata*) sleeping in a laboratory were played recordings of their song; the birds responded by producing the same neural patterns that operated when the birds were awake and singing those songs (Dave and Margoliash 2000).

Other tissues may also be restored during sleep. In young animals, sleep is when the central nervous system matures. In adults as well, growth hormones released during sleep promote the synthesis of protein and RNA (ribonucleic acid, which creates proteins necessary for some cellular processes). These hormones also activate lipid metabolism, which synthesizes and degrades lipids in cells, breaking down fat for storage or energy. Testosterone release increases during sleep, improving skeletal muscle protein synthesis and remodeling of bone. Cell membrane repair, intracellular transport, and clearing of metabolic waste through the kidneys are other important functions that take place during sleep. All these decrease energy demands and cellular stress during wakefulness, so that animals can in those hours perform necessary functions unique to wakefulness (Schmidt 2014).

Phylogeny of Sleep

Animals of every phylum that has been studied for sleep behavior have shown the basic features of a period of immobility during which they have been less responsive to stimulation. Even animals without a brain, such as jellyfish, have a regular rest period where they are harder to stimulate. In animals that have a central nervous system, either a brain or a cephalic ganglion, the definitive signs of sleep are still more pronounced. Flatworms spend their time either swimming or in contraction, where they remain immobile for minutes or hours and are less responsive. Experiments with crayfish, scorpions, spiders, honeybees, fruit flies, snails, and octopuses, among many others, have shown similar but varied results. The next stage of sleep research will be to determine whether any form of sleep exists in organisms without neurons, such as sponges, fungi, or even plants. Among animals, the pervasive evidence of sleep indicates that it must have evolved very early to span such a broad range of phyla with otherwise differing nervous systems (Lesku and Ly 2017; Rattenborg and Ungurean 2023).

Curiously, among vertebrates the recognizable aspects of sleep are less consistent. This is due in part to the definitive features ascribed to sleep having originally come from research on mammals in which measures of neuronal activity in the brain using electroencephalography were the hallmarks. In fishes, for which such measures would be difficult to record in water, presence of sleep has been evaluated by behavior. Some fishes are continually active in all light conditions, while others have definite periods of immobility. This has been treated as "behavioral sleep." In studies with nine species of frogs and salamanders, most showed periods of behavioral sleep, when the animals were less responsive, but experiments with an electrograph, measuring the electric activity in the brain, showed patterns different from those of sleeping mammals. Similarly, among sixteen turtles and tortoises, crocodilians, lizards, and snakes studied, electroencephalogram characteristics vary widely and differ from those of mammals. Most tortoises and terrestrial turtles show behavioral sleep, while sea turtles do not; this may be because, like marine mammals, sea turtles must ensure they are on the water surface whenever they need to breath. Various monitored crocodilians, lizards, and snakes slept three to twenty-two hours per day, indicating a wide range of needs and perhaps of functions (Campbell and Tobler 1984).

Among the distinctive evolutionary features of sleep in birds is that it occurs in much shorter bouts than in other classes of animals, often running only one to four minutes, with frequent brief wakeups that enable birds to check for any danger around them. Most research demonstrating this, however, has been done in laboratories, where birds are usually kept singly, may be fed *ad libitum*, and are free from signs of danger, all of which may affect sleep duration. Recent experiments with wild Chinstrap Penguins (*Pygoscelis antarcticus*) in Antarctica have shown that sleep can even be a matter of seconds, done frequently throughout the day, and adding up to what for humans would constitute "a good night's rest." Using remote sensing monitors, scientists found that during December, when living in constant daylight, penguins slept more than ten thousand times a day, for an average of four seconds at a time. This, however, added up to more than eleven hours of the twenty-four. The birds slept this way both while they were at sea and in their nesting colony. At sea, they need to be alert for predators, while in their colony they are constantly exposed to egg thieves and aggression from neighboring penguins (Libourel et al. 2023).

A defining feature of sleep in birds and mammals is two distinct forms of electric activity in the brain: slow-wave sleep (SWS), or quiet sleep, and rapid eye movement sleep (REM), or active sleep. In birds, sleep bouts with both these features are much shorter than in mammals. Within birds' typical sleep bouts of one to four minutes, periods of SWS last around fifty seconds during the first hours of sleep, decreasing to about twenty-five seconds in the final hours. REM is less than 10 percent of total sleep time, occurring throughout and generally increasing later (Vorster and Born 2015). REM occurs in periods of two to ten seconds, in clusters, often of hundreds in each sleep session, much more briefly than in mammals, where REM runs uninterrupted for minutes or tens of minutes.

The proportion of sleep time in REM is a mean of 8 percent in birds and 17 percent in mammals. REM sleep, however, varies substantially among birds. It is less than 5 percent of total sleep time in European Starlings (*Sturnus vulgaris*), Rooks (*Corvus frugilegus*), Budgerigars (*Melopsittacus undulatus*), and Turtle Doves (*Streptopelia turtur*), but 16 percent in White-crowned Sparrows (*Zonotrichia leucophrys*) and 25 percent in Zebra Finches (van Hasselt et al. 2020). All these birds, however, were tested in laboratories, which did not replicate aspects of seasonal conditions and reproductive states that may influence extent of REM over the course of the year.

The two sleep states were first measured in birds by connecting electrodes on the brain to a computer via a cable tether, which restricts the bird's movements and may affect normal behavior. Today, laboratory research uses wireless technology that transmits electroencephalographic (EEG) signals to a receiver, leaving the bird unrestrained. And now, lightweight EEG data loggers can be used with animals larger than 100 grams. This has created opportunities to study sleep directly in at least some birds in the wild, especially when data loggers are paired with video cameras (Aulsebook et al. 2016).

During SWS, EEG readings show low-frequency, high-amplitude activity. SWS originates in the hyperpallium. The pallium is the equivalent of the mammalian neocortex, but it is structurally different. The slow oscillations during SWS support communication between different, sometimes distant, brain regions and integrate processing of information. SWS is also is linked to synaptic downscaling and renormalization that balances synaptic connectivity throughout the brain, preparing the neuronal network for encoding new information in the next wake phase (Vorster and Born 2015).

In REM, EEG activity is similar to that during wakefulness, with low-amplitude, high-frequency activity. For this reason, REM has sometimes been called paradoxical sleep. Some of REM's visible signs are the closing of both eyes, rapid eye movement, occasional bill movements, and, for birds that rest their head on their back, reduced muscle tone (Wellman and Downs 2009). The short duration of each REM sleep, when muscles are relaxed, was once thought to protect perching birds from falling, but it has since also been found in many birds that sleep on the ground (Lesku and Rattenborg 2014). Domestic geese can continue in REM sleep while standing on one leg if their head is resting on their back; if the head is erect, however, muscle tone is higher and the birds are in SWS (Dewasmes et al. 1985).

Body temperature is lowered during REM, which may be a reason for the brevity of each episode as well as the total amount. At the same time, however, the brain warms during REM and cools during SWS (Rattenborg and Ungurean 2023). In both birds and mammals, the time spent in REM sleep increases after sleep deprivation (Lesku et al. 2009). In birds that sleep at night, SWS occurs principally in the first hours of sleep with, in some birds, REM increasing later. Domestic pigeons (*Columba livia*) given twelve-hour periods of darkness slept 81 percent of the time, with REM making up 4 percent of sleep in the first six hours and 8.7 percent in the second six hours, during which sleep overall decreased in the final two hours (Tobler and Borbély 1988). In nature, the relative share of time spent in each form of sleep may depend on local conditions. Geese spend more of their sleep time in SWS when fasting than when well fed; during fasting, each SWS sleep bout is longer and there are more of them. For REM, bouts are briefer and less frequent when fasting (Dewasmes et al. 1984). The same researchers found similar patterns in Emperor Penguins (Dewasmes et al. 1989).

Seeking the origin of SWS and REM, scientists have searched for it in reptiles, but experiments have been inconclusive. Some of the reptiles tested have two forms of sleep, but the patterns of neuronal activity, durations, and timing in the twenty-four-hour cycle are different from those in birds and mammals. Unlike birds and mammals, reptile limbs do not twitch during sleep and brain temperature does not change in different parts of sleep bouts. In addition, these features differ among species. Thus far, modern tests have used lizards. New research with crocodilians, the closest living relatives of birds, may show more clearly how sleep evolved in birds (Rattenborg and Ungurean 2023).

The parts of the brain involved in sleep in birds and mammals, while having the same origin, are now very different structurally, so sleep features may have evolved independently (Rattenborg 2006a). Comparison of sleep in species from the oldest living groups of birds and mammals—ostriches and platypuses, respectively—show similarities not present in species that evolved more recently. During REM sleep in ostriches, the typical avian rapid eye movements, relaxed muscle tone, and closure of both eyes are evident, but electric activity in the forebrain oscillates between REM sleep-like activation and SWS-like slow waves. This is unlike any other birds that have been tested, but it has also been found in platypuses. In addition, both ostriches and platypuses have relatively more REM sleep than other birds and mammals. These shared features suggest a common, but independent, sequence of steps in the evolution of sleep. SWS and REM likely arose from a single state that later became entirely separated into two distinct states, with the activation of the forebrain as a new feature that performs functions not found in reptiles and other more basal animals (Lesku et al. 2011).

In mammals, factors affecting total sleep time as well as proportions in SWS and REM include body mass, relative brain mass, basal metabolic rate, gestation period, and predation risk. In birds, only predation risk has been found relevant. A literature review of sleep duration in twenty-three bird species (including ducks, turkeys, parrots, owls, doves, penguins, and a range of passerines) found that, in natural situations, the relative safety of their roosting site from predators was the sole feature that correlated with total sleep time (Roth et al. 2006). Eurasian Blackbirds (*Turdus merula*), which in nature sleep in leafy vegetation, were found in a laboratory to spend 6.6 percent of total sleep time in REM (Szymczak et al. 1993). Five Little Penguins (*Eudyptula minor*), which sleep in burrows, spent a mean of 16.2 percent of total sleep time in REM (Stahel et al. 1984).

For both birds and mammals, the time spent in sleep increases when more time has been spent awake. The return to the usual balance of sleep—homeostasis—seems to be internally governed and may be linked to the evolution in both birds and mammals of large, heavily interconnected brains that perform complex cognitive functions and need time to restore and maintain optimal interconnectivity and cognition. In birds, the amount of SWS in an extended sleep bout varies in different parts of the brain based on how intensively each region was used during the prior period of wakefulness. Areas more intensively used have lengthier SWS. If, for example, one eye received more visual input during wakefulness, the visual part of the brain that received input from that eye will have more slow waves during SWS (Rattenborg and Ungurean 2023). Pigeons experimentally deprived of sleep during the day had more SWS during the first three hours of recovery sleep, and they had more slow wave activity during REM sleep (Martinez-Gonzalez et al. 2008). Pigeons deprived of sleep for twenty-four hours then slept longer and with more time in REM sleep (Tobler and Borbély 1988).

Unihemispheric Sleep

The relatively short bouts of sleep typical of birds reflect the balance necessary between the need for neural maintenance and the risks that may be increased by a lengthy complete shutdown of the brain. While a complete shutdown might be the safest and most rapid way to perform neural maintenance, it may also make the bird more vulnerable to predation. In addition to sleeping in bouts far briefer than those of mammals, birds have evolved a way to maintain a degree of alertness during these bouts: unihemispheric sleep. In SWS, birds can sleep either with both eyes shut or with one eye open. EEG activity in the brain hemisphere opposite the open eye is then intermediate between wakefulness and SWS, while the hemisphere opposite the closed eye is in SWS.

Unihemispheric sleep is found elsewhere only in some marine mammals, enabling them to surface and breathe while asleep. Whales and dolphins, eared seals, and manatees have each acquired this ability in their separate evolutions from terrestrial mammals—while sea turtles, in which no sleep of any kind has been found, have not.

For birds, the function of unihemispheric sleep is to enable predator detection while not rousing the entire brain and body. That it occurs only in SWS may be because wakefulness in one hemisphere may interfere with REM sleep in the other. If memory processing occurs during REM sleep, that may require both hemispheres. The brevity of REM sleep, when both eyes are shut and muscles are more relaxed than in SWS, is itself an adaptation to reduce vulnerability. Finally, the loss of muscle tone in REM sleep may counteract what benefits might come from unihemispheric sleep in this state (Rattenborg et al. 2000).

Birds control which hemisphere is asleep. Is one hemisphere given more sleep or wakefulness? The brain's left hemisphere specializes in control of patterns of behavior that occur regularly in familiar situations, while the right hemisphere is responsible for detecting and responding to unexpected stimuli in the environment (Rogers et al. 2013, p. 97). In chicks, which acquire these fundamentals in their first weeks of life, the left hemisphere of the brain is sensitive to major changes in the environment, such as the approach of a predator, while the right hemisphere is sensitive to changes in detail. The left hemisphere is also where memories are consolidated and therefore may require more sleep than the right hemisphere. That could lead to more closure of the right eye, at least during early developmental stages, as has been demonstrated in experiments with chicks raised with or without an imprinting object. Those reared with an imprinting object closed their right eye during sleep more often during their first week, but by the second week after hatching, chicks reared with and without an imprinting object both slept with their left eye closed more than the right (Bobbo et al. 2002).

Later, roosting birds may favor whichever eye and opposite hemisphere is most relevant to the immediate situation, eventually ceding sleep to the other so the entire brain's needs can be met (Rattenborg 2000). In experiments with Mallards (*Anas platyrhynchos*) where four birds were placed in a row, the birds at the end of each row spent more time in unihemispheric sleep than did the more central birds, and approximately 86 percent of that time they opened the eye on their exposed side. That these birds were vigilant in unihemispheric sleep was demonstrated by their responses in fractions of a second to a video simulating an attacking predator (Rattenborg et al. 1995).

Stimulus for Sleep

Role of Melatonin

The inclination to sleep is governed by several interacting sites, including the retina of the eye, the hypothalamus in the brain, and the pineal gland. The eyes inform the hypothalamus of changes in light level; soon after darkness, the hypothalamus secretes hormones that stimulate the pituitary gland to produce melatonin, which is synthesized from serotonin by one of its enzymes and is released into the bloodstream. Melatonin communicates with pacemakers in the hypothalamus and depresses the synthesis of an avian neurosteroid (7α-hydroxypregnenolone), only recently identified, that stimulates locomotor activity (Tsutsui et al. 2012). The retina also produces melatonin, but this is not released into the bloodstream in all species. For some, that comes entirely from the pineal gland, which in birds, unlike mammals, is sensitive to light. The importance of the pineal gland was shown by experiments in which this gland was removed. House Sparrows (*Passer domesticus*) thus treated lost their regular rhythm of waking and sleeping. Other species require the elimination of both pineal gland and the eyes to abolish the rhythm of melatonin in the bloodstream and in behavior (Gwinner and Brandstätter 2001).

Removal of the pineal gland is more disruptive for some birds than others. Circadian rhythm is totally abolished in House Sparrows, White-crowned and White-throated Sparrows (*Zonotrichia albicollis*), and House Finches (*Haemorhous mexicanus*), but only partially in European Starlings, and it has has far less effect on chickens and Japanese Quail (*Coturnix japonica*). This may indicate that in some species or orders other factors are involved (Takahishi and Menaker 1980). In experiments where sleep-deprived birds with pineal gland extant are given a twelve-hour infusion of melatonin, they return to normal rhythms. Further demonstrating the power of melatonin, when infusions are given to birds after nocturnal sleep, this stimulates SWS even when the birds are exposed to light (Berger and Phillips 1995).

Under natural conditions, bright light suppresses the synthesis of melatonin and alters or eliminates a bird's normal circadian rhythms of melatonin production, of feeding, and activity. Species living at high latitudes during the season of continuous light produce less melatonin, and they also produce less melatonin during the weeks or months of almost total darkness, when they need to be awake long enough to feed themselves. Similarly, migratory birds that fly at night, and therefore need to be awake more hours than during their breeding season or in winter, then have less melatonin than when fat reserves are too low to enable migration and they are inactive at night (Gwinner and Brandstätter 2001).

In the Rock Ptarmigan (*Lagopus mutus hyperboreus*) on Svalbard, Norway, at 74–81°N, where daylight is continuous between May and August and where from late November to mid-January there is only a short period of civil twilight, melatonin production varies intensely. From May through July, there is no daily rhythm of melatonin production. Around the winter solstice, melatonin production is also reduced, compared to the spring and autumn months. During these months of total darkness, ptarmigans are intermittently active at all hours, but never as much as during the months with daylight. They retain some of the daily rhythm from the light months, with melatonin level lowest at noon and slightly higher during the other hours (Reierth et al. 1999). Similarly, Emperor Penguins (*Aptenodytes forsteri*) at 66°S in Antarctica have a well-defined pattern of melatonin production during spring and autumn, but they produce hardly any during summer or winter (Miché 1991).

Few nocturnal birds have been studied for melatonin rhythms, but the results have been consistently different from those of normally diurnal species. Barn Owls (*Tyto alba*) and Ural Owls (*Strix uralensis*) have low melatonin levels both by day and night and no discernable rhythm. The Swallow-tailed Gull (*Larus furcatus*) of the Galapagos Islands, the world's only fully nocturnal gull, forages at sea on fish and squid during the night, when these are closer to the surface. By day during the breeding season, the Swallow-tailed Gull is on land and intermittently active, preening, displaying, etc., while during the non-breeding months it is continually at sea, mostly between the islands and the South American mainland. It has low melatonin levels throughout the twenty-four-hour cycle, while the similarly sized diurnal Black-headed Gull (*L. ridibundus*) of Eurasia has a clear rhythm with more melatonin in the bloodstream at night (Wikelski et al. 2006).

Light levels also affect the production of other hormones that influence activity. Corticosterone regulates energy, immune reactions, and stress responses. For birds at temperate latitudes, corticosterone level is linked to the hours birds typically sleep: in diurnal species it is higher at night and in nocturnal ones higher during daylight. During the continuous light of the Antarctic summer, however, Adelie Penguins (*Pygoscelis adeliae*) at 64°S have consistently lower corticosterone concentrations (Vleck and Van Hook 2002).

Internal and External Temperature Changes

While melatonin secretions are making a bird drowsy, its body is undergoing another change that promotes sleep: body temperature. Birds routinely lower their body temperature during periods of sleep to conserve energy through the hours they cannot feed and replenish their energy stores. In birds, metabolic heat production is set in the central nervous system by spinal cord temperature, not, as in mammals, by temperature of the hypothalamus. During sleep, spinal thermosensitivity lowers, first during SWS and still more during REM sleep. The change in body temperature may be related to changes in breathing rate, which is slower when sleeping, and in acid-base balance. At the onset of sleep, more carbon dioxide is retained in the body; this is reversed when birds are aroused (Heller 1988).

Ambient temperature also has an influence. A study of Great Tits (*Parus major*) that use nest boxes in southern Germany found that temperature affects the length of each sleep bout, experimentally confirming what had already been noted, that birds in warmer climates waken more frequently. Under natural February conditions, the birds woke up 4.5 times per hour. When the temperature in the nest boxes was raised experimentally by 5°C, they awakened 30 percent more frequently, an additional 1.3 awakenings per hour, with the greatest effect in the first part of the night, when more time is in SWS than REM. The birds kept under normal conditions also woke more frequently in the first part of the night than later. Frequent wakenings during these hours may be because the first part of the night is when nocturnal predators are most active and birds can maintain more vigilance in SWS. As the night progresses, the drive to restore sleep balance increases and overrides vigilance behavior (Stuber et al. 2017).

Emperor Penguins, during their fasts of several weeks while incubating during the Antarctic winter, increase the time they spend sleeping and thereby lower their metabolic rate while reducing body temperature only slightly in the extreme cold (Groscolas 1990). Similarly, Dark-rumped Petrels (*Pterodroma phaeopygia*) nesting on high, cold mountains in Hawaii, with incubation bouts of eight to twenty-three days before they are relieved by their mate and can return to the sea to feed, spend about 95 percent of this time asleep, during which their respiratory rate falls from twenty-four to twelve breaths per minute (Simons 1985).

Effects of Sleep Deprivation

The duration and intensity of a bird's activity influences the need for rest and the time given to it. To restore sleep balance, homeostasis, after particularly intense or lengthy activity, birds may sleep longer or more deeply. In a Herring Gull (*Larus argentatus*) colony in Cumbria, England, adults that were provided food at their nest remained sedentary and slept less than neighbors that had to fly and search for food (Shaffery et al. 1985a). Glaucous-winged Gulls (*L. glaucescens*) at a colony in Puget Sound slept more during the most demanding stages of feeding their young, starting when these were twenty days old and required more foraging commutes to a garbage dump; the gulls then slept 61.3–93.6 percent more at night than in the earlier stages of chick-rearing (Shaffery et al. 1985b). Great Frigatebirds (*Fregata minor*) returning to their colony from several days spent entirely in flight over the ocean sleep longer and with more time in REM sleep, where muscles are most relaxed, than do ones that have remained at the colony (Rattenborg 2017). Among Chinstrap Penguins in Antarctica, which were found to sleep in bouts of seconds when incubating, those returning from more than twenty hours at sea spent more time in SWS during their first two hours on land than during later periods of sleep (Libourel et al. 2023). Barnacle Geese (*Branta leucopsis*) experimentally deprived of four or eight hours of sleep during the night in summer and winter showed different seasonal responses: in summer, they made up for lost sleep with an increase in sleep the following day. By the end of the twenty-four-hour cycle, they had slept almost as much as the control geese, which were not disturbed; during winter there was no response. Evidently, at least for geese, the need for homeostasis varies at different times during the annual cycle (van Hasselt et al. 2020).

Sleep deprivation has been found to affect immune function, neurogenesis in the adult brain, consolidation of recently acquired information, ability to acquire new information, and alertness, among other features (Roth et al. 2010). Birds have evolved ways to restore homeostasis after deprivation. Laboratory experiments giving birds varying amounts of sleep deprivation have shown similar results in different species. California White-crowned Sparrows deprived of six hours of sleep and then monitored over twelve hours of recovery showed that total SWS did not increase compared with their normal sleep. Instead, the SWS was more intense than normal, as measured by the increase in low-frequency EEG, known as slow wave action. REM sleep decreased during the first two hours; through the recovery period, REM bouts overall were briefer than in normal sleep (Jones et al. 2008).

Pigeons have been used to reveal further refinements of slow wave action: during sleep it may increase only in the parts of the brain that need extra restoration during sleep. Pigeons that were kept awake watching David Attenborough's *The Life of Birds* with only one eye were then found to have increased slow wave action only in the part of the hyperpallium (which processes visual information) that is neurologically connected to the eye that had been stimulated. As in unihemispheric sleep, this was the side of the brain opposite the open eye. For these birds, there was also more REM sleep in the final hours of sleep afterward (Lesku et al. 2011).

Yawning

Yawning is associated with the onset of sleep and the initial stages of waking, as well as with other transitions such as when rest is interrupted, at the initiation of a new phase of activity, and with a reduction in stress level. It can cool the brain, thereby improving alertness and mental processing efficiency. Together with stretching, yawning is widespread in birds and mammals, frequently following one another or occurring together. Yawning is preceded by a rise in temperature in the brain and skull, where it drops after yawning. Experiments with Budgerigars showed that they are more inclined to yawn when ambient temperatures are rising, especially at higher levels, than when temperatures are descending.

Yawning is found both in social and in nonsocial animals and in seclusion, indicating it has a physiological function. In some birds, yawning is contagious. The benefit of contagious yawning may be that by cooling the brain it improves group vigilance. Budgerigars seeing neighbors yawn then themselves yawned more often than otherwise, but in similar experiments Common Ravens (*Corvus corax*) did not respond to others yawning (Gallup 2022).

A study of yawning behavior in Ostriches (*Struthio camelus*), where this can be observed more easily than in smaller or less continually conspicuous birds, found that chicks yawned as soon as they had recovered from the effort of hatching and ridding themselves of the eggshell. The yawn was in the same form as used through all later stages of life. When older, nestlings walking a few paces from their nest yawn, stretch, squat, and settle for a nap. Then and as adults, they awaken with a deep yawn associated with neck stretching, followed or preceded by a complete stretch. For Ostriches, yawning also has a social function connected with rest or sleep. At the beginning of a period of rest for a herd of Ostriches, higher-ranking birds are the first to yawn, indicating absence of danger. Contagious yawning then induces a general relaxation of tension and triggers sleepiness in the group. After any disturbance during a rest, yawning by dominant birds assures more nervous ones that the danger has passed and stimulates them to continue their rest or sleep. At the end of a period of rest, yawning stirs up the group, initiating and synchronizing the new activity cycle (Sauer and Sauer 1967).

Circadian Rhythms

For all birds, perhaps for all animals, presence or absence of light is the principal factor in circadian rhythms, the daily pattern of activity and rest. The morning's increase in light intensity is a positive signal to diurnal species and a negative one to nocturnal ones, with different species beginning or adjusting their activities in response to different light levels, sometimes associated also with temperature. Light level is not directly correlated with duration in time. At most latitudes, light level and extent move forward and backward on the clock with the earth's rotation over the course of the year. At all latitudes, even near the equator, where day length hardly changes, local conditions, such as variations in cloud cover, affect light level and extent on a daily basis. Finally, hormonal levels, which may be triggered by changes in day length, also influence the circadian rhythm, as in shifting activity patterns during the breeding and nonbreeding seasons.

The time birds sleep during the twenty-four-hour cycle depends on the rest of their ecology as well as on seasonal changes governing light level. Compared with winter, birds sleep less during the breeding season if this is at a latitude of longer day length. The season of longest days correlates, for most birds, with the breeding season, so parents that feed or guard their young are active more hours than when they are tending only themselves.

Under Long Hours of Light

In the high Arctic summer when light is continuous, male Pectoral Sandpipers (*Calidris melanotos*) at Point Barrow, Alaska (71°N), sleep hardly at all during the three-week period when each is competing to mate with as many females as possible. The males defend a territory, sometimes with physical fights against other males, and display in flight and on the ground for females, chasing them as well—all energetically demanding activities. A study with marked individuals found that the males slept, in many short bouts, between 2.4 and 7.7 hours each day. The males that slept least sired the most offspring and were most likely to return to the same territory the following year. Later in the season, since the males have no role in incubation or raising the young, their sleep increased to the same levels as in females. For at least a few weeks, however, minimal sleep did nothing to reduce fitness; in fact, it enhanced it (Lesku et al. 2012).

Also at Point Barrow, Semipalmated Sandpipers (*C. pusilla*), in which both sexes incubate, members of a pair replace each other at consistent intervals, but these vary from pair to pair, ranging from twenty-one to nearly twenty-eight hours. For them, the twenty-four-hour circadian cycle is abandoned for at least the weeks of continuous light. Blood samples taken from these sandpipers and also from Red Phalaropes (*Phalaropus fulicarius*) and Snow Buntings (*Plectrophenax nivalis*) sharing this habitat found no detectable levels of melatonin at any hour (Steiger et al. 2013). At 73°N on Bylot Island, Canada, Snow Geese (*Chen caerulescens*) at the beginning of their nesting season rest only 8 percent of the twenty-four-hour day and this in short bouts at any hour (Gauthier and Tardif 1991). Still higher, at Spitsbergen, 77–80°N, Northern Fulmars (*Fulmarus glacialis*) feed at all hours during summer, but more rest on water between 9 p.m. and 6 a.m. The amount of time each fulmar rests, however, is not known (Cullen 1954).

In Antarctica as well, the continuous summer daylight affects sleep patterns. At a colony of King Penguins (*Aptenodytes patagonicus*) on the Crozet Peninsula, birds returning from or departing for the sea spend time sleeping in a resting area away from their breeding territory. They stand close to one another, with at least two penguins per square meter. The depth of their sleep varies at different hours. At all hours, the walkers prefer to pass near sleepers to avoid being attacked by awakened birds defending their territory. Birds sleep more deeply during the afternoon than in the morning, when more birds are walking about. The morning traffic rouses more neighbors from sleep than the lower-activity afternoons (Dewasmes and Loos 2002).

At lower latitudes in Alaska, Rufous Hummingbirds (*Selasphorus rufus*) require less than all the 20.3 hours of available daylight for feeding, courtship, and nesting. Their total resting time, based on observations that could not distinguish quiet wakefulness from actual sleep, was 4.7–5.7 hours per day. Wintering in Jalisco, Mexico, where midwinter daylight is 11.4 hours, they feed for 10.5–10.9 hours; during the actual night, they are presumed to be sleeping (Calder 1993). Similarly, in Churchill, Manitoba, where summer daylight runs twenty hours, American Tree Sparrows (*Spizella arborea*) are active for less than eighteen hours (Baumgartner 1937). Experiments keeping high-latitude Pine Grosbeaks (*Pinicola enucleator*) and Common Redpolls (*Acanthis flammea*) at constant levels of low light found that they shifted their sleep phase from a single period to two or three (Palmgren 1949).

Seasonal Shifts

Birds living year-round at one latitude also have seasonal sleep patterns. As with migratory species, among sedentary birds there is no absolute amount of time they require for sleep. European Starlings in Germany sleep 7.5 hours per day during summer and 12.5 hours in winter, and in summer take naps at midday. Experiments with starlings kept in seminatural conditions showed that REM sleep averaged only 1.3 percent of total annual sleep time and was most extensive during winter nights (van Hasselt et al. 2020).

Blue Tits (*Cyanistes caeruleus*) in southern Germany sleep 4.8 hours longer in winter than in spring. From November through January, they begin to sleep after sunset; in February and March somewhat before sunset; and in April, when the days were much longer and they were feeding young, again at sunset. Throughout the study period, the birds woke up each morning after the start of civil twilight but before sunrise. Individuals monitored through the night woke up between sleep bouts 23 to 230 times, but this did not change seasonally (Steinmeyer et al. 2010).

A comparison of several British tits found that, throughout the year, the smallest species were the first to leave their roost site in the morning and the last to enter it at the end of the day (Perrins 1979, p. 123). This is probably because they need more time to feed to maintain their high metabolic rate, since heat dissipates faster from small bodies than larger ones. In winter, when foraging hours are most reduced and energetic demands are highest, the smallest birds must make the most of all the time they have to find food. Small birds like tits, which are active through all the daylight hours because their food source and foraging method require it, are more likely to have a single, long sleep session when they cannot forage.

Influence of Food Availability

This contrasts with birds such as raptors and carcass scavengers that are active only intermittently during the day, because their food comes in large packages or unlimited quantities. Raptors and scavengers— vultures, storks, and some gulls—may spend hours of the day perched and immobile, in a mix of quiet wakefulness and actual sleep. At Lake Naivasha, the African Fish Eagle (*Haliaeetus vocifer*), which hunts only by day, spends 94 percent of daylight hours perched, and the night as well; actual fishing time is often less than eight minutes per day (Brown 1980, p. 53). A similar condition can be created in the laboratory: pigeons with unlimited food slept 37.7 percent of a twelve-hour light period and 81.6 percent of a twelve-hour dark period (Tobler and Borbély 1988). Likewise, birds that have to travel far from their nest to forage but capture large items may also spend much of the day resting or sleeping. During the breeding season, Northern Gannets (*Morus bassanus*) spend 50 percent of their time asleep while at or near their nest at intervals during the long daylight hours and through most of the night (Mowbray 2002).

Shorebirds in tidal environments, especially those that probe for food rather than searching visually and therefore can forage at night, have a sleep regime governed by tides rather than the sun. They are often polyphasic sleepers, with discontinuous rest periods. Just as with the birds that sleep less during the seasons of longer daylight—including many of the same Arctic-nesting shorebirds—so they may during other seasons on an almost daily basis adjust their hours of sleep in response to feeding opportunities made possible by the tides. Over the course of the year, the extent, timing, and regularity of their sleep changes in response to local conditions. For shorebirds, and likely for other birds as well, age and the experience that comes with it may also influence resting time. Among Bristle-thighed Curlews (*Numenius tahitiensis*) wintering on Laysan Island, adults spend more than 80 percent of daylight standing still, sleeping, or preening, while those in their first year spend 61 percent of daylight performing these activities (Marks et al. 2002).

Sleep Postures

Head and Neck

Most of the nonlaboratory information on bird sleep comes from knowing the characteristic postures birds assume in different stages of sleep and how these differ from the ways birds stand, sit, perch, or float while awake. Many birds begin sleep with the neck relaxed and the head facing forward, sunk between the shoulders. This is the most vigilant posture for most birds; at any distance, they may look awake to a potential predator, even more so if they are sleeping with one eye open. Some birds, including grebes, storks, hornbills, and pigeons, always sleep this way (Thomson 1964, p. 710). Hummingbirds do as well, with the bill tilted upward (Skutch 1973, p. 58).

Many other birds, when feeling secure, then turn their neck so that the head rests on the back, often with the bill buried in the scapular feathers. Birds can maintain both the head-back and the head-forward positions when they are standing, sitting, perched, or on water. The head-back posture reduces heat loss from the head and neck, but is a more vulnerable position, even when the bird has an eye open. When geese sleep with their head facing forward, their muscle tone is higher than when the head rests on the back. They may be in SWS or REM when in either position, but when facing forward in REM, the head drops. Even when the head is supported on the back, and neck muscles are relaxed, however, REM is very brief, with more than 90 percent of REM episodes lasting less than ten seconds, and REM comprising only 10.2 percent of total sleep time (Dewasmes et al. 1985). Some birds, including Eurasian Blackbirds only turn their head back when sleeping in darkness, which may likewise be when they are least vigilant (Szymczak et al. 1983).

The evolutionary antiquity of sleeping with the head turned to rest on the back has been demonstrated by two recent, separate discoveries of fossils of a newly described species of small non-avian dinosaur, *Mei long*, in the same position. They were found in the lower Yixian Formation in Liaoning Province, China (*mei* is Chinese for "to sleep soundly," *long* means dragon). Each is about 53 cm in length, the same size as *Archaeopteryx lithographicus*, the most ancient of known fossil birds, and was preserved in volcanic ash or mud of an early Cretaceous formation 128–139 million years old. Both sit on their hind limbs, with their forelimbs directed rearward (as are bird wings), the neck curved back, and the head resting on the side of the body. While *Mei long* is not an ancestor of birds, the posture with the body and limbs compressed suggests that it too was warm-blooded and had the head against the back to reduce heat loss (Xu and Norell 2004; Gao et al. 2012).

Birds that normally perch by clinging to vertical surfaces sleep with the head forward or back. Woodpeckers, with stiff tail feathers, use these as well as their feet to cling to a tree trunk or the wall of a cavity; several species have been observed with their head back among their shoulder feathers (Villard 1999). Among passerines, the Rock Wren (*Salpinctes obsoletus*) is known to roost on vertical rock faces, propped up by its tail (Brewer 2001, p. 109). Other songbirds living in similar environments, or like tree-creepers and the Neotropical woodcreepers feeding on the trunks of trees, may do so as well. Carolina Parakeets (*Conuropsis carolinensis*) observed by Alexander Wilson and John James

Audubon slept in tree cavities, clinging vertically, with their bill hooked into the cavity wall and feet pressed against it (Snyder and Russell 2002). Swifts, which have relatively short necks, sleep with the head facing forward. Those with stiff tail feathers use these to brace themselves, while the Black Swift (*Cypseloides niger*), which, unlike the rest of its congeners and many other swifts, lacks very stiffened rachises, holds its body away from the rock wall on which it rests (Marin and Stiles 1992).

Most of the birds that sleep on water, such as waterfowl and gulls, also sleep on land, and use the same familiar positions—albeit they are less likely to turn their head to their back if in water with waves or any turbulence. Loons, which normally never set foot on land except on their nest, also sleep with the head on their back. Grebes, similarly aquatic, have distinctive resting postures. They face forward, tucking their bill between the feathers at the base of the neck, most often to the right. One foot, usually on the side where the head rests, may be tucked under the wing to conserve heat. Grebe chicks may turn their head backward, which may recall the way the neck was bent when still in the egg (Fjeldså 2004, p. 78).

Legs and Feet

A posture unique to birds is sleeping on one leg. Birds in many families do it, including some small songbirds, but unipedal standing, when awake or asleep, is best known in storks, flamingos, waterfowl, and shorebirds. It is also seen frequently in herons, hawks, gulls, and owls. Sleeping waterfowl standing on one leg rest their head on the opposite shoulder, while flamingos put the head on the same side as the supporting leg. The benefits or functions of this posture may include reduced muscle fatigue (by shifting weight from one leg to the other) and heat retention, especially if the retracted leg and foot are held among the belly feathers (Clark 1973).

Observation of a flock of captive Caribbean Flamingos (*Phoenicopterus ruber*) at the Philadelphia Zoo showed that they stood unipedally more often at lower temperatures. Conversely, flamingos in the wild in extremely hot environments will lose more body heat with both legs exposed. Another benefit of unipedalism for wild flamingos living in extremely salty or alkaline water is that keeping one leg off the ground some of the time may reduce parasitism and general tissue exposure. The Philadelphia birds were bipedal in windier conditions, because that increased their stability. There was no difference in the length of time resting in one position or the other, suggesting that, for these birds, muscle fatigue from bipedalism was not a factor (Bouchard and Anderson 2011). Resting on one leg is not fatiguing either, which is important if this posture is used during sleep. Specialized structures of the flamingo's limb lock the knee and hip of the supporting leg without any muscular activity (Chang and Ting 2017).

It is likely that a similar mechanism operates for other long-legged birds that often rest unipedally. The thermal benefits have been demonstrated for shorebirds. On the coast of southern Victoria, Australia, observation of nine species from four families, a mix of Eurasian winterers and locals, showed that the longer-legged species roosted more often on one leg than the smaller ones, and six of the nine increased unipedal roosting to over half the time as temperatures decreased. For the smallest and lightest in weight of the sandpipers in these aggregations, balancing on one leg may be more difficult if there is any wind than it is for larger, heavier species (Ryeland et al. 2019). When the temperature is 30°C or above, however, some shorebirds shift to bipedal standing, and those with long bills, such as Dunlin (*Calidris alpina*) also hold the head forward so that heat can be lost through the exposed bill as well as through both legs (Playà-Montmany et al. 2023).

Passerines also sometimes sleep while standing on one leg. British finches, with many cold months to every year, sleep with their head turned back and tucked beneath the shoulder feathers, their body feathers fluffed, and often with one leg raised into the feathers (Newton 1973, p. 130). A Eurasian Blackbird fitted with EEG monitors was on one leg 68.2 percent of the time it was in REM sleep, and 43.9 percent of that time the bird had its head on its back (Szymczak et al. 1993).

Other birds with long legs have several distinctive sleep positions. Standing Painted Storks (*Mycteria leucocephala*) of India and Southeast

Ostriches in brief episodes of REM sleep lose all musculature coordination in the head and neck, which then come to lie on the ground.

Asia often rest on one leg and clasp the tarsus of that leg with the other foot (Hancock et al. 1992, p. 56). Ostriches and rheas sometimes sit with their legs underneath them and their head erect. Other times, they may extend their legs behind them. When their head is erect, ostriches are in SWS, sometimes with one eye open, giving the impression of alertness and maintaining some visual processing. In the transition to REM sleep, both eyes close and the head falls forward, sometimes to the ground. At the end of a REM bout—which, unlike in other birds, may last up to five minutes—muscle tone is restored and they raise their head (Lesku et al. 2011). The Greater Rhea (*Rhea americana*) similarly sometimes holds its head erect or extends its neck on the ground while sleeping, but more often it folds the neck in a tight S-curve, with the nape resting on the back or the base of the neck (Davies 2002, p. 199).

Storks and plovers, which rest most of the time while standing, will also half-squat, putting their weight on their tibio-tarsal joints (Campbell and Lack 1985, p. 104). Among the birds that squat fully are the New World vultures, harriers (which sometimes lie down), and the Secretary-bird (*Sagittarius serpentarius*). All of these except the vultures bury their head in the scapulars when in deep sleep. For the vultures, with their unfeathered head and neck, keeping the head forward may be a way to prevent any parasites acquired when feeding on carrion from reaching feathered areas—the unfeathered head and neck of both New World and most Old World vultures in itself being an adaptation to reduce parasite load (Brown and Amadon 1968, p. 46). Among Old World vultures, the Bearded Vulture or Lammergeier (*Gypaetus barbatus*) sleeps while resting on its belly, with the feet covered by feathers and the head and neck hunched; it does not lie down (Mundy et al. 1992, p. 209). Young birds of many kinds squat by resting on their breast, and adults may do so as well when feeling particularly free from danger; this is seen most often in gallinaceous birds, gulls, and babblers (Thomson 1964, p. 285).

Perching

The general understanding is that birds can sleep while grasping a perch because the stretching of tendons between the ankle and the toes when the leg is flexed causes the toes to curl tightly, with no muscular effort. The more the leg muscles relax, as in REM sleep, the more tightly the toes grip the perch. This has not been tested experimentally, however, and close examination of the legs in some sleeping perching birds shows that this is not always so. Sleeping European Starlings only slightly flex the knees and ankles, and the distal two-thirds of their toes are not flexed to grip a perch. The birds are then balancing their weight over the central pad of their feet—unlike when awake and the toes actively grasp a perch. Starlings that had the relevant tendons or the locking mechanism removed surgically were not able to flex their toes perched in the same manner when sleeping. In other experiments, starlings that inhaled an ether anesthetic were not able to remain perched, indicating that the locking mechanism prompted by the stretched tendon was not in fact automatic (Galton and Shepherd 2012).

Sleeping Blue-crowned Hanging Parrots blend in with the foliage around them at the tips of slender branches, where they are already harder to reach by predators and can easily fly off if disturbed.

Whether or not the toes are actually grasping a perch tightly, on many birds the toes have another feature that keeps a perched bird steady, awake or asleep. Birds that routinely grasp twigs or similar structures usually have a rougher underside of the toes than birds that normally stand on flat surfaces. This is true not just in "perching birds," the Passeriformes, but others as well. Among the ground-dwelling tinamous of the Neotropics, the species that roost on the ground have smooth undersides to the tarsi while those that ascend to trees to sleep have rough ones. There, they do not grip a branch but balance on the rough soles of their feet (Davies 2002, p. 76). As Theodore Roosevelt recounted it, when William Beebe reared the seeds found in the rough tarsi of a Great Tinamou (*Tinamus major*) from Guyana, they were all from arboreal plants (Beebe et al. 1917, p. xi).

Some small parrots sleep while perched hanging upside down. These include the genus *Loriculus*, known in English as the hanging parrots, with ten species ranging from India to the Philippines and New Guinea, as well as the African lovebird genus *Agapornis*, and several species in Neotropical genera. Hanging upside down requires circulatory and musculature adjustments; these may initially have evolved to enable the birds to reach fruit and nectar while clinging head downward. When the ancestral forms could sustain this position—sleeping among leaves at the end of slender twigs that could not support their weight when erect—it would make them inconspicuous and possibly unreachable by predators. When sleeping at night, the head and neck of *Loriculus* are pulled back into the body, and one foot may be tucked into body feathers. When these parrots rest by day, they are upright, with eyes partly closed and feathers slightly fluffed (Buckley 1968).

Fluffing the Feathers

Since birds lower their body temperature at night, they often fluff their feathers to insulate themselves without expending energy. This augments the energy saving some birds are receiving by covering their bill in the scapular feathers, inserting one leg among the belly feathers, or sitting so that both legs are covered. For birds roosting on the ground in open areas, a different strategy is common. These birds seek a slight depression that insulates them on their sides while they sleek down their feathers to make themselves level with the surface and as inconspicuous as possible. Birds roosting in open areas, whether trying to be less visible or not, face into the wind to avoid having their feathers ruffled and therefore less insulating.

Fluffing the body feathers can serve other functions as well, especially in regions where significant heat loss is not a risk. Alexander Skutch (1996, p. 73) observed several species of antshrikes and antbirds in Panama on their nest at night spread their feathers loosely, thereby obliterating their body form; this could be to prevent nocturnal predators from recognizing the shape of the nest or the bird in it. In Southeast Asia, the Fairy Pitta (*Pitta nympha*) roosts on the ground, standing on one leg and putting its head in the fluffed feathers of the back, looking like a ball, which may not match any predator's search image of a bird (Erritzoe and Erritzoe 1998, p. 138).

The Hawaiian honeycreepers are an exception to the usual passerine sleeping position of tucking the head in the back feathers and crouching on their feet to cover them or lifting one of them to conceal it. Relatively warm night temperatures and an absence of biting insects and most nocturnal predators may have led to the loss of habits that their mainland ancestors likely had. As a result, when *Culex* mosquitoes and avian malaria arrived on the islands in the nineteenth century, the honeycreepers were especially vulnerable. The extinction of many species and the severe reduction in the populations of the surviving species have been attributed to avian malaria. Introduced Japanese White-eyes (*Zosterops japonica*), which sleep in the typical passerine position, are only rarely bitten by mosquitoes, while the honeycreepers are bitten extensively. At the time this analysis was made, in 1968, honeycreepers were found on the islands only above 600 m, the upper limit of mosquitoes (Warner 1968). More recently, some species have developed a degree of immunity and have repopulated lower elevations—while mosquitoes have advanced to higher elevations, where avian malaria has decimated honeycreepers there. One species, the Hawai'i 'Amakihi (*Hemignathus virens*), roosts singly rather than in flocks and sleeps with its head tucked behind its back and one foot raised; these habits may have helped reduce its vulnerability (Lindsey et al. 1998).

Eyes

Finally, a few nocturnal birds have eye features used when resting or sleeping during the day. These may deter predators or reduce their conspicuousness. The Northern Pygmy-Owl (*Glaucidium gnoma*) has a whitish lower eyelid that it raises when sleeping, giving the appearance that the owl is awake (Holt and Petersen 2000). Potoos have especially large eyes, most with a bright yellow iris; their upper eyelids have two or three small folds on their edge, creating notches or slits. When the eyes are closed, these small gaps still enable the bird to detect motion. This so-called magic eye may help camouflage the bird, allowing it some vision while resting or in unihemispheric sleep without keeping the entire eye open as would most other birds (Cohn-Haft 1999, p. 289).

Summary

All animals that have been tested—from jellyfish, mollusks, and insects, to birds and mammals—are known to sleep. Sleep's seeming universality suggests that it arose very early in animal evolution. Its key features include a period of immobility, a characteristic posture, and lower response to external stimuli. Sleep must therefore have benefits that compensate for this vulnerable state. In birds, these include consolidating memory, clearing waste from the brain, and restoring other tissues and bones. It is also a time when they can lower their body temperature, thereby reducing energy expenditure during the period when darkness or other conditions prevent them from finding food to maintain their high metabolic rate.

Avian sleep has many distinctive features. Notable among them is sleep's brevity, typically in bouts of several seconds or a few minutes. In between, birds wake up and may check for any sign of danger then immediately revert to sleep. Two patterns of brain activity have been found in sleeping birds: slow-wave sleep, when electroencephalogram readings show low-frequency, high-amplitude activity, and rapid eye movement sleep, when brain waves are similar to wakefulness, with low-amplitude, high-frequency activity. Duration of each REM episode is much shorter than those of SWS, with muscles more relaxed. Over the hours of sleep, the relative proportion of REM increases. When birds are deprived of sleep, they have more REM sleep at their next opportunity, indicating REM's vital restorative functions. Another avian feature is the capacity to sleep with either one or both halves of the brain shut down. Unihemispheric sleep enables birds to maintain a degree of alertness. The eye opposite the awake part of the brain is open. Unihemispheric sleep occurs only in SWS; the greater relaxation of both brain and muscles in REM makes this impossible and may be among the reasons REM is usually so much briefer than SWS.

The inclination to sleep is governed by several interacting sites, including the retina of the eye, the hypothalamus in the brain, and the pineal gland. Soon after darkness, the hypothalamus secretes hormones that stimulate the pituitary gland to produce melatonin, which is released in the bloodstream. This leads to drowsiness and sleep. Under natural conditions, bright light suppresses the synthesis of melatonin and alters birds' circadian rhythms. Species living at high latitudes in seasons of continuous light then produce less melatonin and are active during most of those hours. Birds that remain during the months of substantial darkness also produce less melatonin because they need to be awake long enough to feed themselves. Species that move among latitudes and light regimes over the course of the year usually sleep different amounts depending on the length of the days and nights. Similarly, birds may sleep less during the season when they are busy feeding young than when they are only caring for themselves.

In birds, the most widespread sleep posture is with the neck relaxed and the head facing forward. This is the most vigilant posture; at a distance, birds may look awake, especially if they are sleeping with one eye open. Some birds later turn their neck so the head rests against the back, often with the bill buried in the scapular feathers. This is a more vulnerable position, but the bird loses less heat from the head and the unfeathered eyes and bill. Birds that use these postures can do so when standing, sitting, perched, or in the water. Other birds that normally perch by clinging to vertical surfaces sleep this way as well. A few species squat, stretch out and lie down, or hang upside down. Birds large and small sometimes sleep standing on one leg; this reduces heat loss from the leg tucked under body feathers and may also reduce muscle fatigue by shifting weight from one leg to the other. There is evidently no extra fatigue caused by standing on one leg. Many birds also fluff up their feathers, thereby creating additional insulation or breaking up their body outline, making them less discernible to predators. Despite the widespread understanding that birds sleeping on a perch are grasping it securely (because the stretching of the tendon between the ankle and the toes causes the grip to tighten as the tarsus is flexed), close examination shows that some perched birds are simply balancing their weight on the central pad of their feet.

2

Lighter Rest and Deeper Sleep

In addition to the bouts of slow-wave sleep (SWS) and rapid eye movement sleep (REM) sleep that may occur in shorter or longer stretches when birds have little reason (or ability) to be active, most birds have the opportunity during the average day to rest without slipping into a state that reduces their alertness. Daytime resting for diurnal birds is widespread: the sixteen volumes of the *Handbook of the Birds of the World* note resting behavior in at least forty-nine families, from cassowaries to wood warblers, cranes to larks, etc.

At the other end of the scale, environmental conditions and meta-bolic needs have led some birds to evolve forms of rest deeper than typical sleep, during which they are less alert to danger and harder to rouse. For them, the survival benefits of hypothermia, in which body temperature is lowered more than in normal sleep, outweigh the risks. Torpor, akin to mammalian hibernation, carries hypothermia to a still profounder somnolence that can last a few hours or several days.

Rest

Just as birds may sleep with one eye open, it is likely that birds some-times rest with both eyes shut. Except when using equipment to measure EEG rates, it is impossible to draw a firm line between rest and sleep in wild, free-ranging birds. And since sleep can occur in such brief bouts, it is likely that birds regularly move between the two.

The time available or required for rest depends on many factors, in-cluding weather, diet, and a bird's other daily obligations. Birdwatchers know that spotting birds is most difficult in midday, when birds move and vocalize less. They may be resting between their morning and late-day foraging. Midday in the tropics and in warm weather elsewhere is especially quiet, as birds wait for temperatures to drop from their peak. Some birds need very little time each day to secure all the food they need, and they have most of the day available for resting. Other diets require more time for digestion, during which birds must rest.

Rest time may vary over the annual cycle. Birds defending territories, incubating, or caring for young have less time to rest than birds with no responsibility except survival. During the nonbreeding season when penguins are at sea, they feed in diving bouts with only brief pauses in between the dives; they then rest on the water surface for up to several hours. This contrasts with their pattern when also foraging for their young: Gentoo Penguins (*Pygoscelis papua*) sometimes dive continuously for fifteen hours, with rests of no more than ten minutes, and Adelie Penguins (*P. adeliae*) with small chicks forage for up to eighteen hours without resting (Williams 1995, pp. 94–95).

In species that live across a latitudinal range that gives more hours of light to some populations than to others, birds at different latitudes have different amounts of rest. Swainson's Thrushes (*Catharus ustulatus*) breeding in northwestern Canada at 57°N, where at the summer solstice there are 3.67 hours of complete darkness, are inactive then. At 61°N, where civil twilight (when the sun is between 6–12° below the horizon) persists all through the night, thrushes continue foraging and bringing food to their young in the nest. The thrushes at the higher latitude do not take advantage of the extra hours to increase their feeding rate; instead, they spread the same amount of nest provisioning across the longer time available to them (Ball et al. 2011).

Midday Rest

HEAT

While some birds can spend much of the day resting solely because they have met the day's feeding needs in the morning, more birds seem to rest as a respite from various weather conditions, heat most of all. In nearly all parts of the world, midday heat during the warmer months drives birds to take shelter somewhere cooler and safe—a form of what scientists call "behavioral thermoregulation" to distinguish it from a physiological process such as panting. In moderate ambient temperatures, the average body temperature for birds is 41°C, but this rises in extreme heat. The lethal temperature limit for birds is 46–47°C, an atmospheric level that occurs regularly in deserts and other arid places. In addition, birds have the highest rate of evaporative water loss relative to their body mass of all terrestrial animals (Williams and Tieleman 2005). The combination of these two features makes birds vulnerable to dehydration and energy loss from high temperatures well below their upper limit.

Most birds inhabit landscapes that offer shade, and this is the commonest resort for resting birds to avoid heat exposure. In wooded environments, canopy feeders are the most exposed to direct sun during the hottest hours, so these birds, especially in the tropics, are often the first to withdraw to shade. Other tropical birds with heat-absorbing dark plumage that feed in the open, like many blackbirds and starlings, may be particularly sensitive. In Puerto Rico, Yellow-shouldered Blackbirds (*Agelaius xanthomus*), Shiny Cowbirds (*Molothrus bonariensis*), and Greater Antillean Grackles (*Quiscalus niger*) all retreat, individually or in flocks, to thick vegetation beneath the canopy layer of trees. Yellow-shouldered Blackbirds experimentally exposed to more than three minutes of midday sun usually died (Post and Post 1987).

Nearly every parrot that has been studied is known to rest for a few hours. In the tropics, they seek shade. White-fronted Amazons (*Amazona albifrons*) in Mexico are typical: they spend the morning in flocks, feeding on seeds, flowers, and buds in trees; they then break into smaller groups until late afternoon, when the small groups again coalesce into the larger flock (Levinson 1980). In Bolivia, flocks of Red-fronted Macaws (*Ara rubrogenys*) spend five to seven hours in trees between their morning and afternoon feeding sessions. Sitting or resting occupies as much as 64 percent of that time, with the balance spent preening themselves or their mate, or in what appears as playing or fighting. During some of the resting time, the birds assume the sleep posture with eyes closed and head backward over the shoulder with bill tucked into the back feathers (Pitter and Christiansen 1997). On the treeless Antipodes Islands southeast of New Zealand at 49°S, where sun may be sought more than shade, the two local parrots—the Antipodes Green Parakeet (*Cyanoramphus unicolor*) and the Red-fronted Parakeet (*C. novaezelandiae*)—share their tussock grass habitat with penguins; they both spend much time in the middle of the day basking and preening in sheltered sites on or near the ground (Taylor 1985).

Tropical forest birds that inhabit the several stratified layers below the canopy are not as directly exposed to the sun, but they also move to cooler, more shaded places around midday, when they become less active. In a southern Venezuela rainforest, where midday temperatures are 5–10°C higher than in the morning or evening, birds in mixed-species flocks and those foraging independently were consistently found to move downward—while remaining within their preferred stratum—to denser vegetation as well as closer to tree trunks, which must provide some shade, during the hours when heat is greatest and sunlight most intense (Walther 2002). In the same forest, the numbers of canopy feeders in a fruiting fig also diminished during the hot afternoon, with a short rise again before dusk (Walther 2000). In a lowland rainforest in southeastern Peru, mixed-species flocks moving through trees in the first several meters off the ground were quieter and less active between 12 and 3 p.m., with some individuals preening and resting while others continued foraging (Munn and Terborgh 1979).

Similarly, birds of open country seek shade where they can find it. Northern Wheatears (*Oenanthe oenanthe*) wintering in Africa can experience air temperatures of 40°C. In the desert of Sudan, the surface of the unshaded sand on which they walk can be 80°C, and they may spend as many as six hours of daylight resting in the shade (Moreau 1972, p. 253). Black Eagles (*Aquila verreauxii*) in the rocky landscape of Zimbabwe rest in deep shade during the hottest hours; their alternative to cooling in shade would be soaring to high altitudes with a lower air temperature, but this would expose them to the full glare of the sun on their black plumage, which absorbs heat (Gargett 1990, p. 54). In contrast, Cape Griffons (*Gyps coprotheres*) in South Africa, mostly pale buff colored, take advantage of thermals to soar to great heights to cool themselves, especially after feeding (Mundy et al. 1992, p. 81).

Some birds retreat to the coolness of their nests, even outside the breeding season. Sociable Weavers (*Philetairus socius*) build large communal nests of grass in trees that may hold the separate chambers of hundreds of pairs. In the Kalahari Desert of Botswana, they breed during the cooler winter months, but they roost in their chambers throughout the year; in summer, flocks of weavers return from wherever they have been feeding to spend the hours between 10 a.m. and 2 p.m. in the nest chambers (Maclean 1973). On the high plains of North America, Burrowing Owls (*Athene cunicularia*), which hunt during the day, retreat into the burrows they use year-round when summer temperatures exceed 40°C (Voous 1988, p. 195). Early in the nineteenth century, both Alexander Wilson and Thomas Nuttall noted that flocks of Carolina Parakeets (*Conuropsis carolinensis*) roosted together in tree cavities in the middle of the day during both warm and inclement weather (Snyder and Russell 2002). Alexander Skutch (1989a, p. 186) found that the

Northern Wheatears wintering in Africa seek out shaded places to rest during the hottest hours of midday, when temperatures in the sun can kill small birds exposed for long periods.

female Scarlet-rumped Tanagers (*Ramphocelus passerinii*) nesting around his house in Costa Rica returned around noon to their nests, sitting quietly or sleeping, for more than a hundred minutes, which is much longer than they incubated continuously at other times of day.

During this rest period, some birds drift in and out of sleep, and others are more active. Preening—and mutual preening in many parrots as well as in other bird pairs that stand side by side—is the most frequent activity. Sulphur-crested Cockatoos (*Cacatua galerita*) in Australia sheltering in trees also nibble leaves and strip away bark (Forshaw 1989, p. 141). Many species of mynas and starlings, roosting in flocks at midday, usually in trees, sing, just as they do in their nighttime roosts, indicating that many of the birds are awake at any given moment (Feare and Craig 1999, pp. 152, 188, 219).

Birds resting in hot weather often hold their mandibles apart and pant, which enhances the release of water from the respiratory tract, the avian alternative to mammalian sweating. This may be what Alfred Russel Wallace (1869, p. 568) observed when on a Dutch steamer bound from New Guinea to Batavia, Java, that had a Twelve-wired Bird of Paradise (*Seleucidis melanoleuca*) on board. It had, he wrote, "the curious habit of resting at noon with the bill pointed vertically upward."

COLD, WIND, AND RAIN

"Bad" weather that makes foraging difficult or impossible leaves many birds with no option but to be still and to conserve energy until they can set forth. Even during the severest periods of winter, however, most birds have enough fat reserves to tide them over. A Eurasian Blackbird (*Turdus merula*) can survive 4.6 days without food if it is inactive (Biebach 1977). Smaller birds survive less time, but large birds like raptors that do not always eat every day even in normal weather have no problem as long as they are not soaked and chilled. On September 27, 1985, when Hurricane Gloria passed through the Chesapeake Bay area with winds of 100 km/h, Bald Eagles (*Haliaeetus leucocephalus*) simply spent the day at their roost, waiting it out (Buehler 2000). During very strong winds on the Tasman Sea, wintering Pomarine Jaegers (*Stercorarius pomarinus*) rest on the ocean, facing into the wind and ducking their heads as oncoming waves hit them (Barton 1982).

Rain impedes some birds more than others. Anhingas (*Anhinga anhinga*), which, like cormorants, have fully wettable plumage, will spend most of a rainy day standing motionless and erect with the neck and head stretched upward, enabling water to glide off the feathers (Frederick and Siegel-Causey 2000). In the tropics, some cavity-nesters such as woodpeckers and wrens, retreat to their nests during heavy rains, while other birds find temporary shelter under broad leaves (Skutch 1989, pp. 186–87). Similarly, during periods of inclement weather Verdins (*Auriparus flaviceps*) in the American southwest retire to the well-insulated individual nests they build specifically for winter use (Webster 1999). In rain and in cold, groups of Montezuma Quail (*Cyrtonyx montezumae*) rest together, huddled in tight groups; during rain, these groups sometimes have sudden bursts of mock fighting and short flights (Stromberg 2000).

During winter, birds that normally roost on the ground will take advantage of snow's insulation to make burrows for themselves by diving into soft drifts or digging. While these are typically used at night, birds will also spend the day resting in snow if the weather is so cold that they would consume more energy than they gain by searching for food. During February 1943, when, for a few days in Williamsburg, Massachusetts, the temperature never rose above −28°C, a flock of about a hundred Snow Buntings (*Plectrophenax nivalis*) huddled in their snow holes for most of three days, leaving only occasionally to feed (Bagg 1943). Ruffed Grouse (*Bonasa umbellus*), which routinely roost in snow at night, are perhaps the largest North American bird to rest this way during the day (Rusch et al. 2000).

Influence of Diet

SOME DIETS ALLOW MORE TIME FOR REST

Raptors and fishers often do not need to spend much of the day pursuing food. They are sustained by one or a few meals a day. Even when securing those meals is energetically taxing, their foraging method leaves them many unoccupied hours, sometimes the entire day. The African Bat Hawk (*Machaerhamphus alcinus*) hunts for about a half hour each day, at dusk, when bats are first active; the rest of the day it sits immobile in a large tree, seldom flying (Brown and Amadon 1968, p. 231).

Large birds that eat large fish have plenty of time for rest, especially during the nonbreeding season. A Brown Pelican (*Pelecanus occidentalis*) on the California coast tracked with a radio transmitter for 68.8 daylight hours was inactive for 68 percent of those hours, as well as through the nights, making a total of 81 percent inactive hours (Croll et al. 1986). European Great White Pelicans (*P. onocrotalus*) may catch all their food for the day in an hour or two of the early morning. They spend most of the day loafing, a collective term for preening, digesting, and generally resting. Similarly, cormorants spend 90 percent of the twenty-four-hour cycle either sleeping or loafing (Nelson 2005, pp. 117, 169). During winter, Grey Herons (*Ardea cinerea*) at a day roost in Yorkshire, England, spent 76.8 percent of their time standing but awake, and 5.9 percent in what was taken to be sleep (Birkhead 1973). In Belgium, Grey Herons at a day roost rested an average of 67 percent of the time, likely sleeping an additional 13 percent of the time, mostly in the middle of the day, but the birds at the center of the roost slept 26.7 percent of the time while those at the periphery only 10 percent (Draulens and van Vessem 1996).

Scavengers such as vultures may spend many hours aloft scanning for carcasses, but they do not need to eat every day. When they find a carcass, they can usually feed until satiety. They may return to an unfinished carcass on subsequent days; then, no hours are required for search, leaving more for rest. After they feed, African vultures fly to water sources where they drink and bathe, and they may spend hours standing about and sunning themselves. Their usual hours at water are 11 a.m. until 4 p.m.—ending long before sunset, as the heavy-bodied vultures must get aloft while there are still thermals that can carry them to their night roosts, which may be distant. On still, windless days, they must be even more careful: after bathing, a wet vulture cannot take off; White-backed Vultures (*Pseudogyps africanus*) lie down on hot sand and spread their wings to dry, but while loafing they remain alert with eyes open for jackals and other predators (Mundy et al., pp. 284, 127).

Wintering Yellow-bellied Sapsuckers (*Sphyrapicus varius*) create an analogous situation. They drill well-holes when they arrive in autumn; these usually provide sap all through the winter. Sapsuckers wintering in Texas spend more of their mornings foraging and moving about and 50 percent of their afternoons resting. The amount of rest time increases over the course of the winter, since the well-holes provide sap all through the generally mild Texas winter, especially on warm days. On colder days, when less sap is flowing, the sapsuckers have less time for rest (Speights and Conway 2009).

SOME DIETS REQUIRE MORE REST FOR DIGESTION

In addition to food availability and amounts per effort, digestibility can also influence birds' resting or sleeping patterns during the day. Many birds seem to need time to digest before resuming foraging or any other activity. Rest periods in midday may serve this function as well as keeping birds cool, but for some it is required at other hours, or more frequently during the day. This has been observed in species consuming meat, fish, shellfish, fruit, seeds, and insects.

In Zimbabwe, Black Eagles feed primarily on rock hares and vervet monkeys, also taking some other small mammals, as well as lizards and birds; a pair with bulging crops flew to a perch at 3:10 p.m. on a hot summer afternoon, remained there for nearly 3.5 hours, circled for two minutes and excreted, and returned to the same perch for the night (Gargett 1990, p. 58). White-backed Vultures sometimes doze at a carcass, lying down after feeding (Mundy 1992, p. 126).

Common Eiders (*Somateria mollissima*) wintering in the Gulf of St. Lawrence, Quebec, feed on mollusks, echinoderms, and crustaceans, all of which they swallow whole; the powerful intestines crush the shells and digest the organic matter, but meanwhile the excess weight of the shells not yet excreted makes the birds so heavy that they must rest before feeding again or attempting to fly (Guillemette 1994). Gull-billed Terns (*Gelochelidon nilotica*) wintering on the coast of Guinea-Bissau specialize in fiddler crabs (*Uca tangeri*), which have a low ratio of digestible flesh to exoskeleton; the terns require resting periods to digest and empty the gut before foraging again (Steinen et al. 2008).

Fruit eaters, gaining similarly low nutritional value from the large amounts that must be consumed, also require digesting rest periods. Emerald Toucanets (*Aulacorhynchus prasinus*) in Costa Rica typically visit certain fruiting trees for two to eight minutes, consuming an average of two fruits, each of 13 grams, per visit, and then rest elsewhere for twenty-four to seventy minutes until the seed is regurgitated. In South America, Black-necked Aracaris (*Pteroglossus aracari*) feed primarily on diverse and abundant fruit; after a feeding bout, they may sleep on a branch with their head tucked under a wing, the tail pressed against their back and head, quite as they do when roosting for the night in a tree cavity (Short and Horne 2001, p. 385, 328). In contrast, two species of New Guinea fruit doves sometimes remain in a feeding tree for as long as sixty minutes, but only 31–44 percent of their time is spent feeding; they are loafing for the balance, likely waiting to digest one feeding bout's fruit and excrete its seeds before resuming (Pratt and Stiles 1983).

Gambel's Quail (*Callipepla gambelii*) in the American Southwest feed primarily on seeds, foliage, and a smaller quantity of insects. In Arizona, they typically feed twice each day, in the morning and later afternoon, for a total of three hours on warm days but only forty-five minutes on windy ones. In the middle of the day, the covey retreats to a sandy wash under shade where the birds loaf and swallow sand and grit; this helps them digest the seeds they ate earlier (Brown et al. 1998). The Tufted Jay (*Cyanocorax dickeyi*) of western Mexico consumes invertebrates, fruits, berries, and acorns and lives in small flocks that set off in the morning, traveling as many as 4.8 km in a feeding circuit for 4.5 hours, with intermittent periods of rest; by then, they may have returned to their roosting site or found a similar one where they preen and rest for several hours, resuming feeding later in the afternoon (Crossin 1967). Insect-eating Semipalmated Sandpipers (*Calidris pusilla*) nesting at Point Barrow, Alaska, sleep 10–13 percent of the time during the territorial, pre-laying, and laying periods; sleep is usually after long feeding bouts, suggesting that it facilitates digestion (Ashkenazie and Safriel 1979).

Resting Sites

Birds' varied daytime rest sites are chosen for different reasons. For many birds, the priority is proximity to wherever they feed and clear visibility of any approaching danger, while at night, more birds seek sites that also shelter them from cold, wind, and rain. In tropical regions, birds seek shade at midday; at night they want concealment. Some birds rest in sites completely different from what they use for longer sleep. At night in England, hunting Barn Owls (*Tyto alba*) may stop to rest for an hour or two in farm buildings, such as cattle sheds, that they never use for their longer daytime rest (Taylor 1994, p. 105). Other birds return to their nighttime roost, if this remains the ideal site. Northern Shrikes (*Lanius excubitor*) wintering in Idaho, where they are pursued by hawks and falcons, sometimes return during the day to their night roosts in dense shrubs; these give more protection from wind than the open perches from which they hunt, as well as shelter from predators and an open escape route (Atkinson 1993).

A few birds create specific resting places, while others take advantage of whatever is available. While Ruddy Ducks (*Oxyura jamaicensis*) sleep on water, they sometimes build platforms of vegetation to loaf on, especially during cold periods and when ice thaws (Brua 2001). Black Terns (*Chlidonias niger*) feeding at sea rarely land on water, but they have been seen resting on drifting logs, coconuts, and the backs of basking sea turtles (Dunn and Agro 1995). Black and Bridled Terns (*Onychoprion anaethetus*) at sea also rest on mats of sargassum (Haney 1986). Lesser Black-backed Gulls (*Larus fuscus*) from a colony on the Dutch coast sometimes travel as far as 80 km in the North Sea, where they may rest on the water without feeding for several hours, simply riding the tidal current; some then forage at sea while others return directly to their nest. Why they spend almost four hours flying back and forth to rest far from the colony rather than at their nest, leaving their chicks unattended, is not clear; the gulls may escape ectoparasites while at sea, or restore homeostasis during a period when they can reduce vigilance (Shamoun-Baranes et al. 2011).

Rest site preference may vary individually, even within a flock. European Starlings (*Sturnus vulgaris*) living around English farms, where food is abundant, spend several midday hours resting and preening, with individual birds consistently using two very different sites. Some go to sheltered spots in barns, in thick ivy, or in conifers, protected from wind and invisible to predators, while others from the same flock perch at the top of leafless trees, indifferent to the wind and highly exposed, but ready to fly in a tight group at the sight of an aerial predator. Marking starlings with individualized wing-tags showed that the ones favoring the exposed trees returned day after day to the same tree and even the same branch (Feare 1984, pp. 51, 233).

Resting in Flocks

Social organization also influences the time birds have to rest. Birds that spend their days in flocks have more opportunity to rest if at least some of their neighbors are awake and vigilant. In flocks, the time spent loafing and sleeping may also vary by season, sex, and status. During migration and winter, adult Sandhill Cranes (*Grus canadensis*) spend 50 percent more time during the day than juveniles sleeping or loafing, and males 50 percent more time than females; single and paired birds have more time for these rest activities than parents traveling with their young (Tacha et al. 1992). Alternatively, among flocks of foraging Snow Geese (*Chen caerulescens*), large numbers rest synchronously, in what has been called "infectious behavior" (Mowbray et al. 2000). In the middle of the day, among British finches during the nonbreeding season, feeding flocks break into smaller groups that go off to rest, preen, drink, and bathe. The flocks reassemble in the late afternoon. In midsummer, the break period lasts from about 9 a.m. until 4 p.m., in September from 11 a.m. to 3 p.m., and in midwinter, when the days are shortest and feeding needs are greatest, for an hour around midday (Newton 1973).

Flocks of Dickcissels (*Spiza americana*) wintering in Venezuela enter daytime roosts close to their food source after a few hours of morning feeding. The daytime roosting flocks may number hundreds of thousands of birds. They use isolated trees, forest stands, narrow wooded strips along canals and streams, hedgerows, grasses, and sugar cane, remaining there until late afternoon, when they feed again before returning to their nocturnal roosts. Occasionally, small groups return to these night roost sites, usually fields of sugar cane, after early morning feeding (Basili and Temple 1999). Among small flocks of hatching year Mourning Doves (*Zenaida macroura*) in Alabama, the birds spent 20 percent of their daylight hours feeding and 21 percent resting, mainly at midday; of this resting time, only 3 percent was actual sleep (Losito et al. 1990).

Clubs

Some colonial birds have special communal areas used for resting. Niko Tinbergen coined the term "club" for the sites in the vicinity of seabird colonies that are used by single birds and pairs away from their nests, functioning as a place to rest and perhaps as a place for single birds to find a mate. In the Herring Gull (*Larus argentatus*) colony on the Dutch coast that Tinbergen studied for several years, these were elevated sites with only a flat carpet of dense and very short vegetation—nurtured by the abundant gull droppings—providing a vantage point as well as easy takeoffs, quite different from the topography where gulls put their nests amid dunes and taller grass (Tinbergen 1953, p. 45). Red-legged Kittiwakes (*Rissa brevirostris*), which nest on vertical cliffs on islands in the Bering Sea, have their clubs on horizontal boulder beaches and cliff tops (Byrd and Williams 1993). For Great Skuas (*Catharacta skua*), clubs are used by immature nonbreeders, aged three to seven years, to rest and display. Each club site is usually about 50–100 m in diameter. The presence and size of the clubs reflect the status of their nearby breeding colonies. Newly established and expanding colonies have large clubs, while declining colonies have none. On Foula, west of the Shetlands, small long-established colonies have no club sites; large and densely populated ones have several clubs spaced through the center and around the fringes (Furness 1987, p. 211).

A 1957–59 survey of the seabirds on Ascension Island and adjacent islets in the mid-Atlantic south of the equator found that Masked Booby (*Sula dactylatra*) colonies on Boatswain Bird Island had a hierarchical system of clubs, based primarily on age; Masked Boobies sometimes do not breed until five years old or older. There were six clubs, each at a traditional locale, hosting different demographics: one club usually of adults and near-adults, one of immatures and adults (usually segregated), one mostly of adults and near-adults with a few immatures, another mostly of juveniles of a variety of ages, and the final two exclusively of juveniles and immatures. Mated adults use their clubs when not incubating; immatures use them while elsewhere they were forming pairs. Despite the age segregation of the clubs, tolerated individuals within each club rested more closely to one another than at their nests and even did some mutual preening with adjacent birds irrespective of mating status or age. Curiously, on the same islet, Brown Boobies (*S. leucogaster*) had no club at all (Dorward 1962). At Raine Island, off the Great Barrier Reef in Australia, groups of about two thousand non-nesting Brown Boobies gather on the beaches in long lines two to three deep, facing the sea (Warham 1961). In Northern Gannet (*Morus bassanus*) colonies, only nonbreeders resort to clubs (Mowbray 2002).

Nighttime Rest for Nocturnal Birds

Rest is most easily observed in diurnal birds, but nocturnal ones similarly rest during parts of the night. Many nocturnal birds, in fact, should be more strictly defined as crepuscular, since they are most active during the two daily periods of low light level rather than during total darkness. This gives them a span of inactivity during the darkest hours as well as through the daylit hours. Their activity increases on nights with moonlight and at higher latitudes, where, during some months, there is never complete darkness. Near the equator, where night both comes and ends more abruptly, more nocturnal birds are active in greater darkness. At all latitudes, their schedules are governed by the hours their prey is most active and easily captured.

Without applying individual tracking devices to nocturnal birds, following them is difficult, but for owls there is ample evidence that most are indeed most active in the hours near dusk and dawn. The European Scops Owl (*Otus scops*) is known to rest between midnight and 2 a.m.; the Brown Hawk Owl (*Ninox scutulata*) of southern Asia rests for two to three hours after midnight; the Eurasian Tawny Owl (*Strix aluco*) is most active at the beginning of the night and just before sunrise, but in summer at high latitudes its hunting peak is at midnight and often continues until well after sunrise (Voous 1988, pp. 45, 178, 213).

In the Southern Ocean, where over the course of the year daylight varies from near-continual to almost absent, albatrosses are active in every month, feeding both by day and by night. During the day, prey is easier to spot, but some fish and squid important in albatross diets rise closer to the surface in darkness and are accessible to them only then. As a result, albatrosses seem to have no fixed daily requirements for sleep, rest, or digestion—they are neither strictly diurnal nor nocturnal. Heart rate measurements of Wandering Albatrosses (*Diomedea exulans*) show that their hearts are beating no faster when gliding than when resting on the water or on land. Flapping flight, however, is much more strenuous, so for birds dependent on wind to power their flight, the night hours, generally least windy on the ocean, sometimes force rest (Brooke 2004, p. 8).

Albatrosses can navigate in darkness, but they fly more on moonlit nights. Individuals of four species equipped with satellite transmitters, Global Positioning System loggers, and wet/dry activity loggers were found to spend more time on the water at night, likely a mix of resting or sleeping and "sit-and-wait" feeding. Wandering Albatrosses and Grey-headed Albatrosses (*Thalassarche chrysostoma*) are both consistent in the proportion, but not the amount, of the night they spend on water, be it during the brief summer twilight or the long winter darkness. This results in more hours on the water during the longer nights that offer more profit for the sit-and-wait strategy of feeding, as well as more time potentially available for resting (Phelan et al. 2007).

Nocturnal Activity of "Diurnal" Birds

Many birds that are active during the day are also busy at night, at least during certain seasons. Some are opportunistic foragers, adjusting their schedule to the hours their food is most available or is safest to pursue, while others may forage at night because they did not secure enough

With a full moon, many shorebirds like these Dunlins forgo rest, instead taking advantage of the light and the extreme low tides to increase their feeding hours.

during the day. Migratory birds often travel at night when temperature and aerodynamic conditions (as well as safety from predators) make these hours optimal, recuperating when they land. Highly vocal birds take advantage of night's acoustical properties to sing more effectively than they can during the day. Seasonally or even daily, these birds may shift their resting and sleeping schedules to accommodate other high-priority activities

Shorebirds that depend on the tide to expose their feeding areas may forage at any hour, resting or sleeping during the periods when foraging is least profitable. A yearlong survey of twelve shorebird species at North Humboldt Bay, California, found that the extent of nocturnal foraging varied significantly by season and by species. In autumn, day/

night foraging was observed, respectively, 82 and 64 percent of the time; in winter 100 and 42 percent; and in spring 79 and 14 percent. Some species were active more often at night than others, reflecting their foraging strategies and adaptations. American Avocets (*Recurvirostra americana*), a tactile feeder, and Black-bellied Plovers (*Pluvialis squatarola*) and Semipalmated Plovers (*Charadrius semipalmatus*), which both have relatively large eyes compared to other shorebirds, were the most consistent nocturnal foragers. Among the sandpipers, species that probe into mud rather than locating their prey visually were the most frequent at night (Dodd and Colwell 1996). During December and January on an Atlantic beach in Florida, wintering Sanderlings (*Calidris alba*) continue feeding well after dark, stopping to gather in dense flocks for rest periods of twenty to forty minutes and then resuming feeding (Burger and Gochfeld 1991a). At least eighteen species of gulls are known to be intermittently active at night, taking advantage of fluctuating tidal cycles or seasonal feeding opportunities, such as storm petrels and small alcids that return to their colonies after dark, themselves seeking to reduce predation. Some gulls also copulate during the night (McNeil et al. 1993).

Among waterfowl, species that forage in exposed sites may be safer there at night than by day. Many geese feed in fields at night, but when the moon is bright, they stay in aquatic habitats. For them, as well as for ducks that feed with their head underwater, it is better to be gregarious, alert, and less active during the day on open water where they are less accessible to predators. In Wyoming and Montana, nonbreeding Trumpeter Swans (*Cygnus buccinator*), conspicuous at all hours, congregate in staging areas in flocks of 100–150 where they sleep, feed, preen, swim, fly, and engage in agonistic behavior both day and night. Meanwhile, the breeding pairs are also active at night on their territories; between April and the end of July, males were active on all or part of 83.3 percent of nights observed, females on 50 percent. When feeding at night, swans were regularly accompanied by three species of ducks as well as American Coots (*Fulica americana*), presumably taking advantage of the food the swans stirred up and perhaps also depending on them for vigilance or predator repulsion (Henson and Cooper 1994).

Other birds with ecologies seemingly less suited to full functioning at night also occasionally use the dark hours for feeding rather than resting. The active hours of vultures of the unrelated Old and New Worlds families are generally limited to those when the sun has generated thermals on which they can ride to high altitudes as they search for carrion. Typically, this limits their time aloft to the middle of the day; they remain at their roost longer in the morning than most other birds, and they return to it sooner in the afternoon, when thermals dissipate. At a 45-hectare landfill in Pennsylvania, however, Turkey Vultures (*Cathartes aura*) found a way to extend their day; they continue feeding through the year as much as 2.5 hours after sunset, using the thermals above flared methane vents to rise 200 m before gliding to their nearby roost (Mandel and Bildstein 2007).

From more natural settings, there are also occasional reports of vultures active at night. In Botswana, White-backed Vultures have been seen feeding on an elephant carcass at 9:30 p.m. on a moonlit night (Mundy 1992, p. 130). In alpine pastures in northwestern Spain, as many as fourteen Griffon Vultures (*Gyps fulvus*) were found feeding on two cow carcasses on five nights during one summer. Between sunset and sunrise, individuals were present from thirty-three minutes to six hours. They were recorded by motion-triggered remote cameras that showed no nocturnal vulture activity at ninety-one other carcasses under observation (Mateo-Tomás and Olea 2018). Similarly, of thirty cameras at carcasses in Campeche, Mexico, one recorded a twenty-seven-hour span where during the night hours as many as ten Black Vultures (*Coragyps atratus*) at a time were feeding there, with different birds coming and going; none of the Turkey Vultures or King Vultures (*Sarcoramphus papa*) that had been present during daylight remained. On this night, the moon was two days past full (Charette et al. 2011). Unlike Turkey Vultures, none of these species locate prey by scent. It seems likely that the carcasses were discovered earlier in the day, when some of these birds may have been feeding at them, and they continued into the night because safe roosts were within easy flying distance. Another alternative is that they landed in trees nearby at the end of the afternoon and flew down to the carcass at or after dark when other scavengers had departed.

Several parrots found in Australia, New Guinea, Wallacea, and the Philippines as well as two of the three species in Madagascar are active at night, resuming flight and feeding after they have come to roost. Most feed in treetops, some also on the ground; presumably they can find food in the light available. Observations of most of these species state that they are heard and seen flying on moonlit nights, but some, such as Müller's Parrot (*Tanygnathus sumatranus*) of Sulawesi and the Philippines, are active even on moonless nights (Forshaw 1989, pp. 58, 125, 132, 206, 208, 215, 310, 311).

A survey of 749 North American breeding birds found that nocturnal vocalization has been reported from at least 30 percent of them, across 18 of 22 orders, and 51 of 82 families, of which over 70 percent are considered diurnal. The tallies, however, include birds that are fully nocturnal as well as those that call while migrating at night, in crowded roosts, and in other situations that do not represent an interruption in a period of rest. But during the breeding season many birds of several orders sing during all hours of the night, not just as night turns to dawn, that otherwise are silent through the night the rest of the year. Nights generally have less wind than daylight hours, making for better sound transmission; there is also less acoustical competition, so for some birds it is worth forgoing some rest to vocalize and to be attentive to the responses of others (La 2012).

Some passerines intermittently use the night for more than singing. During the breeding season, Yellow-breasted Chats (*Icteria virens*) sing infrequently at night. Both males and females also then move about within and beyond their territory after dark. Before they have laid eggs, fertile females seek out extrapair copulations; at the same time, males are reducing their own forays to spend more of the night guarding their mate. Using radiotelemetry, the chats in an area containing several breeding pairs were tracked, revealing that both sexes move through the territories of others at night and that they also visited a nearby forest where no chats breed. Such sites may be where chats go to seek extrapair matings. Nearly all this activity ended before the dawn chorus began each morning. As the breeding season progresses, and female fertility wanes, the forays taper off (Ward et al. 2014).

Hypothermia

All birds lower their body temperature when they sleep. For birds that sleep at night, these hours are also usually the coldest, when more of the day's nutrients would be consumed to maintain the high body temperature the bird had when active. And for birds far from the equator, for many months the night is longer than the day—especially at higher latitudes, when these months are also the coldest. For all these reasons, lowering body temperature while asleep is a valuable efficiency. That thermoregulation is suspended entirely during rapid eye movement sleep is likely another reason REM is typically in bouts of only several seconds. Even for large birds, lowering body temperature can be a significant energy saving. Bald Eagles in captivity during winter in northwest Washington state dropped their temperature from 40.7°C to 38.9°C, conserving 4.7 percent of their total heat production, or 4.6 percent of their daily energy budget (Stalmaster and Gessaman 1984).

Hypothermia is a deliberate lowering during sleep of the metabolic rate and body temperature, usually by not more than 10°C from a bird's normal daytime temperature of 40–42°C. Its signs are temporary sluggishness and loss of coordination. Hypothermia consists mainly of slow-wave sleep, when birds have more muscle control than during their brief episodes of rapid eye movement sleep (Heller 1988). If undisturbed, a sleeping bird's body temperature returns to daytime levels before sleep ends, so it is fully functional at the time it normally begins the day's activities.

Hypothermia and torpor, still deeper sleep at a lower body temperature, are best considered as part of a continuum rather than as discrete phenomena. A global review found that hypothermia and torpor are known in birds ranging in size from hummingbirds less than three grams to the Eurasian Griffon Vulture (*Gyps fulvus*) at about 6500 grams. Hypothermia and torpor have been reported in ninety-five species from twenty-nine families representing eleven orders. There are no data from at least 138 additional families, and no data from any birds in the Indo-Malaysian region. From high latitudes to the tropics, from mountains to sea level, some birds routinely drop their body temperature by several degrees for at least a few hours every night (McKechnie and Landgrove 2002).

Benefits and Risks

When birds have found less food than their normal needs, and therefore must stretch the heat-generating power of what they have consumed, they temporarily drop their body temperature more than usual during sleep. In experiments, pigeons deprived of food depress spinal thermosensitivity progressively more each night as their energy reserves decline, with their body temperature showing comparable reductions. Their thermal sensitivity reverts to baseline on the first night after re-feeding. The same pattern has been found in smaller birds (Berger and Phillips 1995). For these, the benefits of hypothermia are even greater. They have likely worked harder during the day to find food and, because of their diminutive size, will dissipate heat far more rapidly.

Hypothermia also has its costs or tradeoffs. While it is used regularly by some birds, it is used by others as a last resort. Before resorting to hypothermia, a bird's first available energy-free tool to maintain a consistent body temperature while asleep is to fluff its feathers to reduce heat loss. A roost site with insulation, like a tree cavity, can also do some of the work that otherwise falls to the bird. Cavities, however, are relatively scarce. Birds that have access to them can further raise the interior's heat level by huddling with others. The Pygmy Nuthatch (*Sitta pygmaea*) is the only bird known thus far to deploy both cavity roosting and huddling in groups within the cavity—as well as hypothermia—to get through the night. Grouped nuthatches have a lower body temperature and lower breathing rate than ones roosting singly (Kingery and Ghalambor 2001).

Regularly hypothermic birds can maintain lower fat reserves, but their reduced response rate during the hours when their body temperature is lowest increases the risk of nocturnal predation. The alternative, to avoid hypothermia, is to gain more fat during the day, but this increases energy costs and the risk of diurnal predation. Balancing these factors, some birds use hypothermia only during severe conditions and in environments where daily foraging is energetically stressful. Typically, these birds have low fat reserves, are in places with low ambient temperatures, and have a high daily variability in foraging success. An additional factor for some of the small, high-latitude birds, such as tits and chickadees that cache food earlier in the winter or earlier in the day, is whether they have depleted their supplies (Pravosudov and Lucas 2000).

Birds in Higher Latitudes

Hypothermia has been studied most extensively in tits. Willow Tits (*Poecile montanus*) and Siberian Tits (*P. cinctus*) kept in the same conditions from January to February in Finland slept in the classic position, with feathers fluffed and heads turned to their back. They had body temperatures 5°C lower when asleep than when awake, and they slept most heavily when the ambient temperature was lowest, at −19°C. As they slept, their weight loss per hour increased with declining air temperature and decreasing body temperature, to as much as 0.10 gram per hour. Their body temperatures rarely dropped below 30°C. This is some 10–20°C higher than in torpid hummingbirds, swifts, or nightjars; in those, body temperature comes close to air temperature, and metabolism is much reduced. These tits, however, living in regions with much lower ambient temperatures, would freeze to death if they lowered their body temperature that much. As soon as the tits were disturbed and woken, their body temperature began rising, slowly at first, then at maximum speed for two to eight minutes. It took the birds ten to twenty-three minutes from the end of sleep to reach their daytime body temperature (Haftorn 1972).

In North America, Mountain Chickadees (*P. gambeli*) and Juniper Titmice (*Baeolophus ridgwayi*) at high elevations in Utah, where even summers are cold, use hypothermia year-round. In summer, they lower their body temperature 4–11°C below its daytime level, in winter 3–9°C. Individual titmice, but not chickadees, with the least body mass at the beginning of the night lower their temperature more than heavier birds. Hypothermia produced an energy saving of 7–50 percent in the chickadees and 10–28 percent in the titmice, with the greatest savings at the highest ambient temperatures (Cooper and Gessaman 2005). Black-capped Chickadees (*P. atricapillus*), with a much wider range across North America, are less consistent. In regions where the night temperature is 10°C, a black-cap can save as much as 75 percent of its nightly energy expenditure by lowering its body temperature 10°C, but less at more extreme temperatures; in Alaska, they do not enter hypothermia even at −50°C. The birds there are at least one gram heavier than those at temperate latitudes and have more insulating feathers (Reinertsen 1983).

In southern Australia, where the average daily winter temperature range is 0–12.7°C, the Silvereye (*Zosterops lateralis*) lives in many habitats, feeding at this season mainly on nectar because insects and fruit are scarce. Weighing 10–12 grams, the same as a chickadee, it has the same metabolic challenges as the small birds in other hemispheres. When sleeping, its body temperature drops by as much as 5.5°C, independent of the ambient temperature, and its metabolic rate decreases by up to 50 percent. At 7°C, a typical winter temperature, this results in an energy saving of 15 percent (Maddocks and Geiser 1997).

Some larger birds also become hypothermic. In Alaska at 66°N, where midwinter air temperature often falls below −40°C and daylight is four to six hours, Canada Jays (*Perisoreus canadensis*) weigh 65–85 grams. They roost between 4 and 5 p.m., and their body temperature drops gradually, an average of as much as 5.5°C over the next twelve to fourteen hours, stabilizing between 5 and 7 a.m. Body temperature rises sharply to the usual daytime level before they become active at 10 a.m. On colder nights, Canada Jays lower their body temperature more than on milder ones (Waite 1991). Going larger still, a Turkey Vulture weighing 2230 grams kept at 15°C in California had a daytime body temperature of 41°C that dropped at night to as low as 34°C, during which time the bird was not responsive to handling (Heath 1962).

Are there any ecological factors these birds of colder regions share? Canada Jays, like many tits, cache food for retrieval during the winter and early spring, while the Turkey Vulture is a scavenger. The birds that provided these data were held in aviaries, with unlimited food, but in nature all these species may sometimes find little or no food during winter, so that the benefits of reducing their metabolic rate during the night can be substantial.

Low-Latitude Birds with Nutrient-Poor Diets

Similarly, consumers of nutrient-poor nectar and fruit cannot store enough energy on their small bodies to get through the night, so they need to reduce their burn rate. Some small birds therefore also use hypothermia to reduce energy expenditure while sleeping.

Sunbirds of Africa and Asia routinely drop their body temperature. Their diet is primarily sucrose, glucose, and fructose, and they need to feed 75 percent of the daylight hours. Nectar and tiny insects do not enable them to maintain their diurnal body temperature, 42°C, through the twelve-hour equatorial nights. At the high elevations where many sunbirds live, the ambient temperature may for many hours be near 10°C. Sunbirds there may lower their body temperature by as much as 17.5°C. Body temperature will begin to rise long before dawn, even when the ambient temperature is still dropping. This indicates that hypothermia is under the bird's own control, not driven by external factors (Cheke 1971). As with the northern birds, for sunbirds the metabolic reduction derived from hypothermia varies with the ambient temperature: for the Tacazze Sunbird (*Nectarinia tacazze*), which lives as high as 4000 m on Mount Kenya near the equator, the reduction is 51–74 percent at 25°C and 43 percent at 5°C. This is typical for the many sunbird species that have now been studied (Cheke and Mann 2001, pp. 29, 249).

The Neotropical manakins, only slightly larger than sunbirds, consume fruit that digests rapidly. They are unable to store food in their gut or fat on their body, so they benefit from becoming hypothermic at night. When the nighttime ambient temperature in Panamanian rainforests at sea level is 19.5°C, Golden-collared Manakins (*Manacus vitellinus*) reduce their body temperature when sleeping by about 10°C, to as low as 30.5°C. At the same time, oxygen consumption declines by 40 percent. These are both substantial energy savings, especially during the last half of the rainy season, August to December, when fruit production is low. Then, manakins must search farther to find even small quantities. On the many days that their final foraging hours are eliminated by four-hour afternoon rains, the birds are all the more depleted at nightfall (Bucher and Worthington 1982).

Torpor

Torpor is a profounder form of hypothermia, with body temperature lowered to below 30°C, sometimes closer to 20°C or even less. Heart, respiratory, and metabolic rates are greatly depressed, coordination is essentially absent, and response to external stimulation is diminished or absent (Calder and King 1974, p. 344). Torpor is known best from three families: hummingbirds, swifts, and nightjars. Hummingbirds are primarily nectivores, with high energy demands and only minimal storage capacity. Swifts and nightjars are aerial insectivores. All three are subject to short-term variability in rainfall and temperature that affect their food supply. For them, torpor is the most efficient way to wait out hours or days when food is unavailable (McKecnhie and Lovegrove 2002). For most birds, torpor is strictly a nightly phenomenon. One nightjar, the Common Poorwill (*Phalaenoptilus nuttallii*), has been found in this state for several days.

Hummingbirds

Among hummingbirds, torpor is deployed by many species from Alaska to the tropics and in many different ecosystems from lowlands to the Andes at 5000 m. Energy savings are greatest when ambient temperatures descend to a hummingbird's minimum body temperature. The savings from lowering their metabolic rate then range between 60–90 percent of energy per hour compared with the bird's normal daytime temperature. The extent of savings increases with the number of hours in torpor. Body temperature, heart rate, and breathing all decrease. In the deepest phase of torpor, hummingbirds have no reaction to external stimuli. Their feet do not grip a perch with a clamp reflex; instead, their claws are closed around the perch. A short time after they are touched, they stretch their wings and give shrill cries not heard at other hours (Krügier et al. 1982).

   In torpor, both heart rate and breathing rate drop substantially. At an air temperature of 32°C, the heart of an active North American Black-chinned Hummingbird (*Archilochus alexandri*) beats 480 times per minute. At 1°C, it beats 1,020 times per minute in an active bird working hard to keep its body temperature at daytime levels. But when torpid, the heart beats only 45–180 times per minute. Similarly, hummingbirds have a waking breathing rate of 245 times per minute at an air temperature of 33°C, and 420 times per minute at 13°C, but when torpid, hummingbirds breathe only sporadically, and some may have periods of apnea (Baltrosser and Russell 2000).

The propensity to enter torpor is also correlated with the ecology of each species. The Rufous Hummingbird (*Selasphorus rufus*) enters torpor in all seasons, but it uses it most during autumn, when this depletes less of the bird's fat stores accumulated for migration (Hiebert 1993). In Arizona and Ecuador, the large, territorial species of hummingbirds that have ready access to food resources sufficient to carry them through the hours of sleep become torpid only under extreme conditions. In the same habitat, nonterritorial hummingbirds and the smaller territorial species enter torpor regularly in normal conditions. Even within a single species, however, individuals that begin the night with greater body mass are less likely to enter torpor than lighter individuals (Shankar et al. 2020).

A study of six Andean hummingbirds at 3800 m in Peru found distinctive patterns in each, with a varying ability to lower body temperature to that of the environment. Their periods of hypothermia ranged from 2.3 hours in the largest species, the 24-gram Giant Hummingbird (*Patagona gigas*), to 12.9 hours in the 6-gram Black Metaltail (*Metallura phoebe*) and the 4.9-gram Bronze-tailed Comet (*Polyonymus caroli*). While the body temperature of the 8.7-gram Sparkling Violetear (*Colibri coruscans*) never descended below 8°C, irrespective of minimal air temperature, the body temperature of the Black Metaltail matched the descent of the ambient temperature by within less than 1°C. It lowered its body temperature to 3.26°C, the lowest currently known for any bird or nonhibernating mammal (Wolf et al. 2020).

Swifts and Swallows

Experiments with swifts suggest that torpor is a response to food deprivation, not a nightly phenomenon. White-throated Swifts (*Aeronautes saxatalis*) from California have a normal daytime body temperature averaging 38.6°C. Individuals kept unfed at 5°C reduced their body temperature to as low as 18°C on the third day; those kept unfed at 20–22°C lowered their body temperature to as little as 23°C on the second day. When fed again, they recovered their normal temperature (Bartholomew et al. 1957). A White-throated Needletail (*Hirundapus caudacutus*) wintering in Australia on a reduced diet for three nights entered torpor within two hours of darkness; on one night its temperature dropped from 38.5°C to 28°C in two hours when the ambient temperature was 25°C. In the mornings, its temperature returned to daytime levels within a few minutes (Pettigrew and Wilson 1985).

Swallows have occasionally been reported in a torpid state. Like swifts, they are aerial insectivores that sometimes face days of cold and/or rain when no insects are flying. The White-backed Swallow (*Cheramoeca leucosterna*) of Australia nests in burrows it excavates in sand banks. During the winter of 1936, sixteen to twenty were found in one burrow, all inert and oblivious to being handled before they were returned to the burrow (Serventy 1970). During a few days of cold weather in Connecticut in April 1983, three Tree Swallows (*Tachycineta bicolor*) were found sheltering in a bluebird box; one was alert, the other two unaware of their observers. Three days later, the box held two dead swallows, both extremely emaciated and weighing only a third of their weight of four days earlier (Stake and Stake 1983). The capacity to enter torpor was confirmed in experiments with House Martins (*Delichon urbica*) from southern Germany that were kept at low temperatures and without food, matching natural conditions the birds routinely encounter during spring and summer. The birds lowered their body temperature to 25.7°C and their metabolic rate to 10–20 percent of normal. At normal weight, House Martins never enter torpor, even at ambient temperatures of less than 5°C (Prinzinger and Siedle 1988).

Nightjars and Their Relations

Among nocturnal aerial insectivores, torpor is known from a few nightjars, in which it has been induced experimentally and found in the wild, and in related families, such as the frogmouths and owlet-nightjars. In

South Africa, the Freckled Nightjar (*Caprimulgus tristigma*) has been found to reduce its body temperature during winter to as low as 12.8°C. As with other nightjars, the frequency, depth, and duration of torpor varies among individuals. After the birds have finished feeding, torpor has not been found to last for more than a few hours in the latter part of the night, sometimes extending into the following morning (McKechnie et al. 2007). In the coldest months of the Australian winter, nighttime body temperature of Tawny Frogmouths (*Podargus strigoides*) drop to 21–26°C for about seven hours. In winter, Australian Owlet-nightjars (*Aegotheles cristatus*) also lower their body temperature, to as far as 19.6°C, but this begins around dawn and lasts for an average of four hours (Holyoak 2001, p. 53).

The only bird species known to have multiday torpid periods is the Common Poorwill. This was first discovered in December 1946, when one was found sleeping and immobile in a rock hollow in the Chuckwalla Mountains, part of the Colorado Desert, in Riverside County, California. The bird was then banded, thereby identifying it as the individual that spent most of that winter and the two following ones in the same crevice (Jaeger 1949). It was not monitored daily, so the duration of its torpid periods is unknown, but more recent observations and experiments in Arizona have shown that poorwills may remain in the same place without foraging for several weeks. Their roosts are typically open to the south or southwest, providing direct solar radiation in the afternoon. That enables the bird to rewarm passively to more than 25°C before reentering torpor at sunset.

Poorwills with roost sites experimentally shaded (and therefore not able to use the sun for passive warming) actively warmed themselves after at least three days of torpor, always on relatively warm days, but they did not leave their site. Since the actively warming birds did not shiver—the widespread mechanism for birds to increase body temperature—their method is still unknown. Some went for almost seven days without rewarming, and one bird that rewarmed itself periodically did not leave its roost for forty-five days (Woods et al. 2019). Other experiments have shown that poorwills can reduce their oxygen consumption by over 90 percent and lower their body temperature to

5°C, the lowest known for any bird until the more recent work with Andean hummingbirds (Csada and Brigham 1992).

Carolina Parakeet

Only one example of torpor is known from a bird of a completely different ecology, and it unfortunately is not available for modern scientific investigation. Carolina Parakeets (*Conuropsis carolinensis*) lived year-round as far north as New York and Illinois, where they routinely remained through cold winters. They fed primarily on fruit, mast, and nuts, which were available in all seasons. Their habit of roosting in flocks in tree cavities gave them some shelter as well as the heat generated by birds massed close together, but perhaps torpor was required also, at least during times of scarcity. Accounts from Indiana and Florida, in 1842 and 1861, respectively, describe the felling of their roost tree, with the several birds in a cavity oblivious to and unharmed by the disturbance to their site. No one was measuring their body temperature, but the unresponsiveness of the birds suggests they were torpid. The Indiana roost contained "dozens" of birds that, when transferred to a cage on a winter day, revived and fed eagerly. The Florida account, where cold is not mentioned and seems less likely a factor, states only that all the birds were "secured" (Butler 1892; Taylor 1862).

Summary

In addition to the hours birds spend in regular sleep, they have many opportunities, and some needs, for additional, lighter rest during the hours they are active. And, in certain environmental situations, some birds resort to deeper forms of sleep—hypothermia and torpor—that are highly efficient ways for them to conserve energy during the coldest hours of the twenty-four-hour cycle.

Most birds do not need every waking moment to feed themselves or, during the breeding season, their young. The species for which food is abundant, easy to find, and in large packages have even more time when they can loaf—remaining alert for any danger, but not in much need to move about. This leisure time may vary by season, sex, and status, with more experienced birds and those not responsible for young having the most. When these free periods are extensive, birds may return to their roosting site, to another place with comparable conditions, or to ones that match the needs of the hour, if these are different from night time roosts. Some colonial seabirds have "clubs," traditional sites where idle birds gather. These clubs may segregate or exclude birds based on their age and breeding condition as well as by the status of the nearest nesting colony.

While some birds can spend much of the day resting solely because they have met the day's feeding needs in their first hours after waking, most birds seem to rest as a respite from various weather conditions, heat especially. Thus, midday, when temperatures are highest, is usually when birds are least active, often taking refuge in the shade or a sheltered, inconspicuous location. They may drift in and out of short bouts of sleep, preen, or, if not solitary, interact quietly with mates or neighbors in a flock. Rain, snow, or extreme cold can limit birds' ability to find food, so for its duration birds likewise often remain inactive. But since most birds can survive without food for a day or more—larger birds much longer—poor weather rarely leads to starvation if birds are not already in poor condition. Other birds rest because they need time to digest, whether from gorging on a carcass or while processing shellfish, fruit, and other foods with low nutritional value that must be consumed in large amounts.

Nocturnal birds also have rest periods, typically in the middle of the night, since, like diurnal species, they usually feed most intensively upon first waking and then before retiring. Their hunting hours may often depend on the hours their nocturnal prey is most active. The night is also sometimes an active period for some birds that have no fixed hours—shorebirds that depend on shifting tidal schedules, waterfowl that are safer feeding at night in certain habitats, and others that can shift their resting or sleeping schedules to take advantage of unusual food sources. Birds that migrate at night may also change both their sleep and rest schedules on the following day.

Birds have evolved several ways to minimize energy loss through the hours they are asleep. All birds lower their body temperature when they sleep. For birds that sleep at night, these hours are also usually the coldest, when more of the day's nutrients would be consumed to maintain the high body temperature the bird had when active. And for birds far from the equator, for many months the night is longer than the day— especially at higher latitudes, when these months are coldest. For all these reasons, lowering body temperature a few degrees while asleep is a valuable efficiency. Hypothermia goes beyond this. It is a deliberate lowering during sleep of the metabolic rate and body temperature, usually by not more than 10°C from a bird's normal daytime temperature of 40–42°C. Its signs are temporary sluggishness and loss of coordination. If undisturbed, a sleeping bird's body temperature returns to daytime levels before sleep ends, so it is fully functional at the time it normally begins the day's activities.

Hypothermia has its costs or tradeoffs. It is used regularly by some birds, mostly small ones, and by others only as a last resort. Regular hypothermy enables birds to maintain lower fat reserves, but their reduced response rate during the hours when their body temperature is lowest increases the risk of nocturnal predation. The alternative, to avoid hypothermia, is to gain more fat during the day, but this increases energy costs and the risk of diurnal predation. Balancing these factors, some birds use hypothermia only during severe conditions and in environments where daily foraging is energetically stressful. Typically, these birds have low fat reserves, are in places with low ambient temperatures, and have a high daily variability in foraging success.

Hypothermia and torpor, still deeper sleep at a lower body temperature, are best considered as part of a continuum rather than as discrete phenomena. Torpor is a profounder form of hypothermia, with body temperature still further from usual daytime levels, descending to below 30°C, sometimes closer to 20°C or even less. Heart, respiratory, and metabolic rates are greatly depressed, coordination is essentially absent, and response to external stimulation is diminished or lacking. Torpor is known best from three families—hummingbirds, swifts, and nightjars.

All three are subject to short-term variability in rainfall and temperature that affect their food supply. In addition, for hummingbirds, their small size and sugary diet make it challenging for them to store enough energy to get through long or cold nights. Torpor is thus the most efficient way to wait out hours or days when food is unavailable. For most birds that enter torpor, it is strictly a nighttime phenomenon. One nightjar, the Common Poorwill, has been found in this state for several days or even weeks.

3

Where to Roost?

Every day, every bird must choose where it will roost. Unlike many aspects of avian behavior that seem driven primarily by instinct, determining where to sleep is a deliberative decision that is made anew daily. Birds have many options that make each potential site more or less attractive for the coming night, depending on time, season, weather, competition, risk of predation, distance to be traveled, and other factors. Birds that migrate or wander must assess the possibilities at places they may visit only once or, in the course of their travels in habitats over different latitudes, that may be quite different from their most recent experience.

For most birds, the most important factors in a roost site are safety, protection from weather, and proximity to food. Territorial species usually include these in their assessment of where to settle for breeding, wintering, or living year-round. Sometimes, however, the other features they need do not provide ideal roosting sites close by, and these must be sought elsewhere. Similarly, migrants may find food resources far from suitable roost sites. Wide-ranging aerial insectivores, marine birds, wetland and tidal flat specialists, scavengers, and other species with food that is patchily distributed in space and time may be feeding far from an ideal roost site. For all of them, selecting a place to spend each night has its challenges and has led to a wide variety of adaptations and behaviors.

Safety from Predation

For many species, safety is the primary factor governing roost site choice because when they are asleep birds are least able to detect approaching predators and may least be able to evade them in darkness, even if fully roused. This priority may require tradeoffs with other features desirable in a roost, such as proximity to a mate, best positioning to defend the territory, and short distance to feeding sites.

Safety in Concealment

Many birds choose sites where vegetation or other features make them harder to discover or to reach by a predator. The importance of concealment is shown by the number of birds that routinely leave their daytime territory each night because it lacks any place the bird feels safe in the dark. Among Wood Thrushes (*Hylocichla mustelina*) breeding in coastal Virginia, some 31 percent of males roost outside their home range, an average of 121.8 m away from the nest on which their mate spends every night. These males find denser concealing vegetation beyond the limits of their own territory. Younger males roost farther from the areas they use during the day, suggesting that these birds have been able to acquire only inferior territories. After nests fledge or fail, the more distant males return and roost near their mates (Jirinec et al. 2016). For them, preventing another male seeking a copulation for a second nesting attempt may outweigh the territorial male's interest in personal safety during the night.

For some birds, nest architecture rather than location makes these unsafe for roosting. South American Rufous Horneros (*Furnarius rufus*) do not sleep in their sturdy oven-shaped nests made of dried mud, built to shelter eggs and young from the elements, because the adult would be trapped by any predator at the entrance. Instead, they seek dense vegetation, which is sometimes scarce in their territory. Individuals lacking a concealed site in their own territory look for one within the territory of a neighbor, but they wait until that bird has retired to its own roost (Skutch 1996, p. 151).

During winter, the tradeoff can be between concealment and warmth. American Robins (*Turdus migratorius*) in eastern Washington state roosting in groves of Douglas fir position themselves 1–2 m away from the trunk on branches with needles, where they are less visible. They lose the protection from wind that a place adjacent the trunk would provide, but they give themselves a much easier escape route (Walsberg and King 1980). A five-year survey of 623 roosts of wintering Northern Saw-whet Owls (*Aegolius acadicus*) in Wisconsin found that these were consistently chosen for concealment, not thermal benefits; nearly all were hidden from above and from most lateral directions, the ways a predator might approach (Swengel and Swengel 1992).

Since most birds that roost in trees and shrubs are least concealed from below, like the Saw-whet Owls, it is likely that their primary safety concern is predatory birds. An unobstructed escape route beneath the perch is most useful for any bird attacked from above or from the side. Reptiles or mammals that approach from below or on the branch would perhaps shake it and wake the sleeper in time for it to move (Skutch 1989a, p. 34). Some birds that build special nests for roosting, not raising young, demonstrate this: the Stripe-crowned Spinetail (*Cranioleuca pyrrhophia*) of South America builds a tubular sleeping nest of twigs and vegetable fibers that looks like a bit of debris caught on a limb. It has two entrances near the bottom, giving the bird an escape route if a predator approaches on the branch and enters through the other hole (Skutch 1996, p. 148).

Birds that roost on the ground have different challenges finding the safest roost site, because they are vulnerable to predators approaching from both the ground and the air. Male Capercaillie (*Tetrao urogallus*) in Norwegian mixed coniferous forest roost under the lowest branches of spruce trees, preferably in old forest where spruces are densest and there is greatest canopy closure, more shrub coverage, and lower visibility along the ground. This gives the birds the greatest concealment, but it reduces their ability to detect an approaching predator and makes escape from any attack more difficult (Finne et al. 2000). In the dry rocky interior of Australia, the White-bellied Plumed Pigeon (*Petrophassa plumifera*) roosts on gentle slopes on or near hill tops, on the lee side of a low shrub or a large stone, creating a slight depression for itself, but never underneath any vegetation or directly against a rock that might impede its flight from a nocturnal predator (Goodwin 1967, p. 189).

Antipredator roosting habits can evolve in response to new dangers. Bicknell's Thrushes (*Catharus bicknelli*) wintering in the Dominican Republic spend the day on territory in broadleaf cloud forest, but at one study site with an adjacent pine forest, most moved there to roost, avoiding the Norway rats (*Rattus norvegicus*) and the highly arboreal black rats (*R. rattus*) that have more difficulty climbing the vertical trunks of the pines. At night, the pine forest is generally 1°C cooler than the broadleaf forest, so the thrushes are not moving there for warmth (Townsend et al. 2009). Rats have been on the island of Hispaniola for approximately five hundred years; how long it took Bicknell's Thrushes to develop the habit of leaving their territory for nocturnal safety is not known, but a more recent similar phenomenon demonstrates that the behavior can evolve in some birds within a few avian generations. The critically endangered Mariana Crow (*Corvus kubaryi*) is found only on the islands of Guam and Rota. On Guam, the principal cause of decline has been predation by the invasive brown tree snake (*Boiga irregularis*), which became established there after World War II. By the 1960s, on some parts of Guam, the crows began roosting communally away from the territory of each family, a previously unknown habit there or on Rota, which at the time and into the 1990s did not have tree snakes. (It does today.) The ecology of this crow does not suggest there is any reason other than the benefit of increased vigilance from birds being together for why the communal roosting habit recently developed (Wiles 1998).

Safety in Numbers

Communal roosting is widespread among birds, ranging in numbers from small groups to flocks of millions, and sometimes of several species together. It has several benefits, including the likelihood that some wakeful vigilant birds will spot any approaching predator, as well as the reduced probability to each bird in the roost of becoming a victim. To further reduce their risks, flock roosters often choose sites such as isolated groves, wetlands, islands, or urban landmarks that are inaccessible to at least some predators or that are in exposed locations that provide early visibility of any predator's approach.

Perhaps the most famous of the latter is the fifty-mile stretch of the Platte River in Nebraska, where, between February and April, some 500,000 Sandhill Cranes (*Grus canadensis*) stop on their way north to feed on the remains of the prior year's corn crop in the farmland on both sides of the river. They roost in the shallow waters of the river's wide channels. There, they are less accessible to predators than if they remained at night in the fields where they forage during the day, as do some Common Cranes (*G. grus*) wintering in Spain. Today, there are few likely dangers along the Platte, but over the millennia that cranes have made this a staging area, wolves and other predators would have been present in the original grasslands. At night in the river, the cranes spend 8.6 percent of their time on alert, but they are alert 14.4 percent during the day in the fields (Sparling and Krapu 1994).

The safest communal roosting sites may vary from day to night. The daytime roosts used by Grey Herons (*Ardea cinerea*) in a Belgian colony of fifty to one hundred pairs are on the ground, near water, and difficult for ground predators to reach, but they are still near foraging areas; these also provide a broad view. For the night, the birds move to trees, where they are less accessible to predators during the hours these would be most difficult to see (Draulans and van Vessem 1986).

Over the course of a day, shorebirds that feed on exposed tidal flats must weigh several factors in their search for a safe roost site, as the times the flats are available for feeding shift by an hour every day, and the places available for the birds to wait out the high tides also change. In Roebuck Bay, on the tropical coast of northwestern Australia, wintering Great Knots (*Calidris tenuirostris*) and Red Knots (*C. canutus*), when obliged by the tide to roost during the day, choose sites closer to their feeding areas, some one to three kilometers away, selecting places where, to avoid overheating, they can stand on wet substrates, which at midday can be 15°C cooler than dry ones. At night, they sometimes fly six to eight more kilometers to sites that are safer, because these are farther from tall cover. The ideal night sites are also surrounded by dry, light-colored sand on which any ground predator would be more visible. Then, safety is a higher priority than cooling, and the birds prefer to stand on dry sand rather than on darker wet sand close by (Rogers et al. 2006a).

Safety in the Air

In some places with a shortage of high-tide roosting sites distant from elevated or vegetated ground that could conceal a predator, shorebirds resort to a more extreme measure known as "over-ocean flocking." For Dunlins (*Calidris alpina*) wintering at Boundary Bay on the southern edge of the Fraser River in British Columbia, the chief predator is Peregrine Falcons (*Falco peregrinus*), which over the winter capture about 14 percent of the birds there. To avoid roosting close to shore where they would be most densely concentrated and most vulnerable to a sudden attack, flocks of Dunlins fly out to sea one or two hours before high tide, returning two to four hours later, when they can land on more distant flats. The energy expenditure of staying aloft is evidently worth making to avoid predation (Dekker et al. 2004). This behavior was rare in the Fraser River estuary before the mid-1990s and became common thereafter, correlated with the increase of Peregrines, which were scarce there in the early 1970s. In the 1990s, wintering Dunlins had less fat than in the 1970s, indicating that in the balance between risks of starvation and predation, the birds are selected for the agility that comes with less weight, which better enables them to evade falcons. To reduce energy expenditure during these flights, the birds move against the wind with fluttering wingbeats, using lateral airflows to stay aloft (Ydenberg et al. 2010).

At one location on the German Wadden Sea, Dunlins abandon their regular roost sites only when extreme high tides leave no place available far from tall saltmarsh vegetation, and only from mid-September through the winter, when Peregrines and Sparrowhawks (*Accipiter nisus*) are present. During the highest tides of late summer, they roost in the saltmarsh vegetation that later in the year could conceal a raptor. In contrast with the birds at Boundary Bay, here the Dunlins spend no more than ninety-nine minutes airborne (Hötker 2000). Similarly, Semipalmated Sandpipers (*Calidris pusilla*) in the Bay of Fundy have changed their roosting behavior in response to the rising population of Peregrines, which were not present there in recent times until 1982. On days when the highest tides cover all the traditional roosting spots, the sandpipers spend a few hours flying in flocks over the sea (Dekker 2011).

Safety on the Sea

Sea ducks that, except for females when incubating, spend their entire lives on open water where they are highly visible, move from nearshore feeding areas to remoter places where darkness and distance conceal them. In the Salish Sea, the inland marine waters formed by British Columbia and peninsular Washington state, wintering Surf Scoters (*Melanitta perspicillata*) travel an average of 3.5 to 6 km between the inshore waters where they feed to deeper and more distant waters where they spend the night in dense rafts, some as large as 2 km long and 500 m wide. The scoters select waters with low tidal currents to avoid drifting toward the shore or farther from their feeding areas. Other duck species as well as loons and grebes may also be present, all beyond the usual swimming range of nocturnally active predators such as mink (*Mustela vison*) and river otters (*Lutra canadensis*) (Hamilton et al. 2022).

For marine birds with fully aquatic predators, the sea itself is not always a safe roosting place. Petrels, especially storm petrels, frequently have foot injuries, including missing toes or an entire limb, probably taken by predatory fish. In the deep waters off the Atlantic coast of southern Africa, Great-winged Petrels (*Pterodroma macroptera*) and Leach's Storm-petrels (*Oceanodroma leucorhoa*) can be seen in bunches, huddling together on the water with no space between the birds, in groups of up to seventeen great-wings and twenty-three Leach's, sometimes mixed in smaller numbers, looking like a single black object. Because these petrels are black on their undersurface, contrasting with the lighter shade of water, they are easily seen from below by fish, sharks, and marine mammals. By bunching together, they may appear to be a more impressive or potentially dangerous creature, and, in any case, the aggregation reduces each bird's chance of being bitten (Camphuysen 2007).

In other oceans, shearwaters and petrels form dense rafts when they are not foraging. About thirty miles off the coast of San Diego on August 31, 1935, one observer estimated many thousands of Black Petrels (*O. melania*) in "great rafts of birds on the water, much as shearwaters raft. They made black patches like kelp flies so close together as to be inseparable to the eye. Some rafts were 100 yards long and several rafts were visible at once." To a potential predator below, the shape of these aggregations would not have resembled any large marine animal, but, as with the petrels off southern Africa, here too the odds are reduced for every bird in the raft (Miller 1936).

Unlike petrels, boobies feed close to land and can take refuge there when that is the safer alternative. At Sand Island in the Johnson Atoll in the central Pacific Ocean, more Red-footed Boobies (*Sula sula*) roost on land during calm weather; at sea, they cannot get aloft if it is windless and they are then more vulnerable to sharks, which come to the surface when the sea is flat (Schreiber and Chovan 1986).

Safety in Urban and Industrial Sites

Many birds have co-opted urban, industrial, or other manmade sites as communal roosts because these are far from their natural predators or give better vantage points. Shorebirds, which routinely seek a flat exposed place to wait out high tides, have found that various manmade sites provide the same benefits. In both the Bay of Fundy and British Columbia, migrant sandpipers have taken to novel inland roosting sites, such as roads, airport runways, the causeway of a ferry terminal, and a gravel pit (Dekker et al. 2011). In the Hawaiian Islands, Pacific Golden-Plovers (*Pluvialis fulva*) took quickly to the roofs of flat buildings for roosting, both by day and by night. In rapidly urbanizing Honolulu, the roof of a school built in 1966 was used nightly (by at least 1973) by as many as 125 plovers. On the windward coast of Oahu, where more marshes and fields remain, the roof of a building at a Marine Corps air station attracted about 150 golden-plovers for loafing and sleeping during the day, but at night these birds went to saline marshes nearby (Johnson and Nakamura 1981). On the northeast coast of Brazil, at the estuary of Paraíba do Norte, migrant and wintering Semipalmated Plovers (*Charadrius semipalmatus*) and Semipalmated Sandpipers use a warehouse roof 2–2.5 km from the nearest feeding areas as a roost during high tides. They are most numerous during the highest tides, with as many as 876 birds of the two species counted. But the total number using the roof is likely higher, because smaller groups of forty to eighty frequently arrive and depart, suggesting turnover among individuals. Much of the land around this estuary has been developed in the last few decades, leaving shorebirds few natural high-tide roosting sites. Scarcity of alternatives probably drives birds to this roof (Cardoso and Zeppelini 2013).

Some manmade sites are also warmer, but this is likely not always their most attractive feature, since artificial sites are widely used in the tropics as well. In Trinidad, Short-tailed Swifts (*Chaetura brachyura*) have taken to nesting and roosting in uncovered vertical manholes that are part of an underground drainage system. The circular entrances of these cylindrical concrete tubes range from flush with the ground to about a meter above it; they are 1–7 m deep, sometimes with water at the bottom. The usual temperature range in them is 26–30.5°C, with humidity consistently at 95 percent, so the swifts are probably not choosing the manholes for thermal benefits. These manholes also hold nests of two species of wasps, which may deter predators. Sometimes as many as 375 swifts roost in a single manhole, and the roost sites change from year to year among the many manholes, independent of which have been used for nesting (Collins 1968).

Jackdaws (*Corvus monedula*) and European Starlings (*Sturnus vulgaris*) have been commuting nightly to cities for many years, perhaps centuries. Even in winter in northern Europe, however, the thermal benefits of a longer flight to a slightly warmer city roost, estimated at 2–3°C, do not often justify the effort. For Jackdaws roosting in town centers in Sweden, Finland, and the Baltic republics, a flight of fifteen minutes or 7.5 km to the roost consumes as much energy as would be saved by spending the night at a site 4°C warmer, while a daily flying time of fifty minutes would require that the city roost be at least 13°C higher than one in the countryside to compensate for the flight loss. The increased safety must be worth this effort (Gyllin et al. 1977).

Common Ravens (*Corvus corax*) also have traditional communal roosts. Some provide no shelter at all from the elements, such as the 6 km of electrical transmission lines with fifteen transmission towers in southwestern Idaho that sometimes host more than 2,100 ravens. These towers are used throughout the year, with peak numbers in late summer and early autumn. The safety of numbers, the inaccessibility of the lines to ground predators, and the visibility of any aerial predators must be attractive features more important, even in winter, than concealment. The ravens prefer the highest portions of the towers, but more roost on the lower portions when winds exceed 7 km per hour (Engel et al. 1992).

Swallows are conspicuous flock roosters during migration and in winter, with some aggregations in the millions. In natural landscapes, their traditional roosting sites are marshes, which perhaps led to the European belief that swallows spent the winter buried in the mud there. The first accounts of swallows using urban sites for large communal roosts come from the twentieth century, but it is possible the habit began earlier, because several species of swallows, like some swifts, have always found suitable nesting sites on buildings. Since at least the 1960s, wintering Barn Swallows (*Hirundo rustica*) have roosted on wires and buildings of a commercial district of Bangkok that is brightly lit all through the night. The numbers grew from 150,000 (in 1964–66) to 200,000–400,000 (in the 1980s). During the same period, the city's expansion eliminated all the reed beds that previously were roosts, but not evidently the places where these swallows forage; some come into the city from 30 km away (Ewins et al. 1991).

In South America, wintering Purple Martins (*Progne subis*) roost in flocks of hundreds of thousands, originally in marshes and in trees on river islands, but in the 1960s some in Brazil began roosting in urban and industrial sites, including an oil refinery in Manaus. The pipes on which they rest are almost too hot for a human to touch and the air nearly unbreathable. Here, the safety may not be exclusively from predators but also from parasites; biting midges, which infest martins throughout the year, are rare or absent at the refinery (Davidar and Morton 1993). In Iquitos, Peru, the Plaza de Armas is an isolated patch of trees at the edge of the commercial part of the city, one block from the Amazon River. In 1976, three hundred to six hundred wintering Southern Martins (*P. modesta*) spent April through July roosting there. In 1977, the plaza was occupied by an estimated 250,000 from April to October. The following April, 250,000 martins arrived, remaining until their roost was disturbed in mid-August; the mayor had the trees pruned of all their branches because some Iquiteños believed the large concentration of dark birds was an ill omen for the future of the city (Oren 1980).

Thermal Factors

Since temperatures are almost always cooler at night than in the day, and since birds routinely lower their body temperature at night to save energy, a significant factor for many birds in their choice of roost site, if not the primary one, is shelter from cold, wind, and rain. At both high latitudes and in the tropics, some birds make very substantial efforts—from constructing a special roosting nest to commuting long distances—to ensure that they have a roost with thermal benefits. Others may find a suitable site very near where they have been feeding. Dunlins wintering on the coast of Mauritania on the Banc d'Arguin move to roost in areas of dry sand during diurnal high tides. There, if the temperature of the sand is warmer than the air, they lie down in the sand; at night, when the sand has cooled and the air is somewhat warmer, they stand (Klaassen 1990).

Some birds shift their roost site each night or during the night, depending on whether their immediate priority is warmth or shelter from rain or wind. Black-billed Magpies (*Pica hudsonia*) use sites that provide a variety of thermal benefits as well as safety. They favor trees over water, reed beds, and thorny bushes, and they roost communally, sometimes a hundred or more, gaining the benefit of numbers. In Denmark, Eurasian Magpies (*P. pica*) usually perch halfway up a tree, but they use lower branches on colder nights and the leeward side when it is windy. In Alberta, magpies use conifers, which also protect them from both rain and wind. They rest on the lowest foliated branches immediately below the canopy. At −20°C, they can save 8 percent of their energy roosting in dense conifers out of the wind (Birkhead 1991, p. 80).

Other large birds need only seek shelter during extreme weather. The Snowy Owl (*Bubo scandiaca*), with its insulating plumage and low metabolic rate, can survive ambient temperatures below the lowest recorded in the Northern Hemisphere; it is probably indifferent to where it spends the long Arctic night (Boxall and Lein 1989). Conversely, nocturnal birds, especially in warm regions, may select daytime roost sites that are cooler. In Australia, the Spotted Nightjar (*Eurostopodus argus*) can withstand temperatures over 40°C, but it seeks partially shaded places on the ground in open woodlands; during the course of the day, it gradually turns to keep its back to the sun (Holyoak 2001, p. 295).

Warmth

Some birds roost in the open, seemingly oblivious to cold. Most birds seek places that provide some warmth, when it is available. Many birds choose foliage, whether broad leaves in trees and shrubs, or conifer branches with dense needles, dense stands of reeds, or exposed surfaces that absorbed heat from the sun during the day and continue to radiate it after dark. They may still need to fluff their feathers, shiver, or reduce their body temperature to conserve energy. In more sheltered sites that also protect from wind and rain, it is sometimes difficult to quantify the extent to which the site provides a benefit against each of the three thermal factors.

For many birds that, during the breeding season, sleep on or in the nest they use for incubating eggs and for sheltering the young, this is also an insulated roost site. Only a few adult birds, however, continue to sleep in the nest after their young have departed—not all nests are strong enough to have withstood the activity of the young or the impacts of weather, and most nests are likely to have acquired parasites that would prey on any bird that remains. So whether or not the nest is still structurally sound or clean, birds that do roost in anything they build for themselves generally seek out or create something new, and usually only to accommodate themselves rather than a mate or a fledged family. Raptors, storks, and other birds that build durable nests to which they add new material each subsequent breeding season tend to migrate or shift to roosting elsewhere until the next year, by which time the accumulated parasites may have died off.

Some small birds seeking a warm spot for the night occupy the old nests of other birds or build nest-like structures of their own as an alternative to cavities (described in chapter 4). In the Southwest, where even summer nights are cold, Verdins (*Auriparus flaviceps*), only 6.9 g, build individual sleeping nests throughout the year; juveniles build their own soon after fledging. Winter nests are larger, thicker, and more insulated than summer nests (Webster 1999). In Arizona, Black-tailed Gnatcatchers (*Polioptila melanura*) use empty Verdin nests; as many as sixteen have been found roosting together in a single nest. Within these nests, on different nights, the temperature was 8.6–30.2°C above the outside temperature (Walsberg 1990).

In wooded habitats, the sites most sheltered from the weather are cavities in trees, natural or made by birds (discussed in detail in chapter 4). Cavities are highly sought, especially by birds that cannot make their own. Usually warmer as well as protected from wind and rain, they also conceal birds from many predators. Woodpeckers provide most of the nonnatural cavities later used by many other birds, while after the breeding season many woodpeckers create new cavities for their own use until the following year. Whether natural or excavated by a bird, these may be a few degrees warmer than outside temperatures, and the heat generated by the roosting bird in this enclosed space further raises the temperature. Great Tits (*Parus major*) in Slovakia, when given a choice in winter of several uninsulated artificial cavities, examined each in the two hours before sunset and chose the ones that were receiving more solar radiation. These were warmer at the time each bird entered its chosen cavity at dusk and remained warmer through the night than the unused cavities (Velky et al. 2010).

In seasonal environments, many sedentary birds that remain on the same territory through the year change their typical roost site, using more sheltered places when it is cold and others with more air circulation when it is warmer. Eastern Screech-Owls (*Otus asio*) in central Texas roost primarily (77 percent) in tree cavities or birdboxes in winter, but, during the rest of the year, they roost 80 percent of the time in foliage (Gehlbach 1994, p. 33). Spotted Owls (*Strix occidentalis*) prefer cool, shady locations in summer, while in cold weather they perch higher in the canopy and closer to the trunk; during rain, they go beneath overhanging branches (Gutiérrez et al. 1995). Blue Grouse (*Dendragapus obscurus*) in Oregon roost in conifers during the winter, when they also

Spotted Owls, like many roosting birds, move to places with thicker overhanging foliage when it rains.

feed on the needles of these trees; in the warmer months, they roost on the ground, where their camouflage may be more effective than in trees (Popper et al. 1996).

Birds that have consistent feeding territories during the day may move to different elevations nearby to roost if these have a better microclimate. In the mountains of Mindanao, Philippines, Johnstone's Lorikeets (*Trichoglossus johnstoniae*) have in some areas an altitudinal range span of 1500 m; in the evening, they move to lower elevations within this range to roost, returning at sunrise to the higher forests (Forshaw 1989, p. 70). On the eastern slopes of Mauna Kea, Hawaii, during autumn, two honeycreepers, the 'Apapane (*Himatione sanguinea*) and the Iiwi (*Vestiaria coccinea*), descend several hundred meters from the forests in which they feed to roost in warmer ones, but at elevations above the nightly fog belt that envelopes still lower parts of the slopes (Ralph and Fancy 1995). In the Swiss Alps, wintering Bramblings (*Fringilla montifringilla*) feed on beech mast in forests as many as 45 km from the birds' roosts in conifers at lower elevations. There, flocks of millions roost together. They descend a few hundred meters, to roosts that are 2°C warmer than the beech forest, but they avoid valley bottoms, where a local inversion zone develops on clear, windless nights, making these elevations colder than higher ones. On a windless night, the benefits of a roost 2°C warmer outweigh the energy costs of a 10 km flight; on a windy night, the savings outweigh a flight of 36 km (Jenni 1991).

For warmth, just as for distance from predators, birds have discovered the benefits of manmade structures. At petrochemical facilities in Great Britain, thousands of European Starlings roost between the pipes of the oil condensers, which are always warm; at one plant where the pipes were allowed to go cold, the starlings left (Marples 1934). In South Africa, Cattle Egrets (*Bubulcus ibis*) at Stellenbosch typically roost in trees over open water or at the margin of water, where it is 1°C warmer than over land. At a roost under observation when the temperature dropped to 0°C, some egrets moved 80 m to trees overhanging the roof of a factory that generated warmth, or onto the roof itself; when the temperature reverted to normal levels the birds returned to their usual trees. The egrets, which can roost in flocks of a few thousand, normally return night after night to the same tree, each bird to the same branch. They may have learned from experience (rather than a random search) where to move for greater warmth (Siegfried 1971).

Wind and Rain

Wind and rain may be regular or intermittent phenomena in any habitat, occurring together or separately. The benefits of reducing exposure to each over the course of a night are substantial. For small birds, protection from wind is sometimes a more critical factor than from cold. Dark-eyed Juncos (*Junco hyemalis*) wintering in Indiana in an area of old fields and deciduous woods roosted in dense evergreen shrubs, where the temperature through the night over the course of a winter was in the same range ($-17°$ to $20°C$) as elsewhere nearby, but the wind speeds were about 81 percent less than those measured outside the roost (Webb and Rogers 1988). Phainopeplas (*Phainopepla nitens*) wintering in Arizona roost in the densest portions of small trees within their territory; each bird usually has two or three such sites. These roosts are not any warmer, but their shelter from wind provides five times more thermal benefit than an exposed site (Walsberg 1986). At a roost of Eurasian Blackbirds (*Turdus merula*) in London, birds sought the thickest part of bushes and trees, where there was a 73 percent reduction in wind speed and therefore a reduced windchill. On nights with a high windchill, birds in the most sheltered spots lost only about 60 percent of the weight lost by birds in more exposed places within the roost (Clement and Hathway 2000, p. 48).

Rain can chill birds—despite their water-resistant plumage—especially when prolonged and at hours when it is too dark for roosting birds to move far to find a more sheltered spot. Combined with cold, which is intensified by wind, the results may be fatal. For Red-winged Blackbirds (*Agelaius phoeniceus*) a wet plumage increases the standard metabolic rate fivefold, and normal thermoregulation cannot be maintained at temperatures below $15°C$ (MacMillan and Carpenter 1980). At a roost of 25,000 blackbirds and European Starlings in Urbana, Illinois, a storm in February 1939 killed about 4 percent of the birds there. Many froze to death when the temperature dropped to $0°C$ during the rain. Common Grackles (*Quiscalus quiscala*) and Brown-headed Cowbirds (*Molothrus ater*) roosting in sheltered conifers nevertheless suffered higher rates of mortality than starlings, seemingly more resistant even in leafless trees (Odum and Pitelka 1939).

As with so much else, cavity roosters have an advantage in protecting themselves from wind and rain. Among the woodpeckers that excavate new cavities in autumn or winter to use until the next breeding season, the Downy Woodpecker (*Picoides pubescens*) orients these away from the prevailing wind, while cavities made during other seasons are randomly oriented (Jackson and Ouellet 2002). In a tropical rainforest in Costa Rica, it was found that four woodpecker species, large and small, orient most of the holes they excavate to the northeast and to open areas. This may increase the cavities' exposure to sun, accelerating evaporation of nightly condensation and of rainfall, as well as allowing the prevailing winds to penetrate (Rico and Sandoval 2014). Cavities used by Mountain Chickadees (*Poecile gambeli*) and Juniper Titmice (*Baeolophus ridgwayi*) in Utah were found to reduce windspeed enough that temperatures were 12.1–13.7°C higher inside; the energy savings could have enabled the birds to go an additional 5.7–7.3 hours without food (Cooper 1999). Red-tailed Amazons (*Amazona brasiliensis*) in southern Brazil roost communally in trees during the cold and windy non-breeding season, but they continue through the year to visit and use the tree cavities in which they nest. On sunless and wet days, however, fewer parrots come to the roost; they have most likely opted to spend the night in their nest cavity (Cougill and Marsden 2004).

In many habitats, protection from wind and rain may be available at the same site, but these may be limited and birds sometimes travel considerable, energetically taxing distances to reach them. Bald Eagles (*Haliaeetus leucocephalus*) likely have no predators to fear wherever they roost, but they are sensitive to weather conditions and choose a protected site. At a river in northwest Washington state, wintering eagles feed on salmon on the gravel bars; at dusk, they leave the bars and ignore the nearby leafless deciduous trees to fly 2.6 km or more to evergreens. Their flight is all by flapping, strenuous over such a long distance for a bird of 4.5 kg, but the greater warmth in the conifers, combined with protection from wind and rain, saves energy during the night. The energetic cost of their typical flight distance is only 30–37 percent of what is saved by roosting in a less exposed situation (Stalmaster and Gessaman 1984).

Some small birds similarly travel far to sites sheltered from wind and rain. These may also provide safety from predators. Northern Waterthrushes (*Parkesia noveboracensis*) wintering in Puerto Rico each have their own daytime territory in several forest types, including some with no standing water. Around sunset, most waterthrushes leave their territory and fly as many as 2 km to stands of red mangrove (*Rhizophora mangle*), where they often return nightly to the same tree, and some the following year as well. There, the waterthrushes, and individuals of three other wintering warbler species, spend the night. The benefits of this site may include the dense canopy of the red mangroves that blocks rain and wind. In addition, the warblers are at the edges of a communal roost of local Gray Kingbirds (*Tyrannus dominicensis*), which are much more alert during the night, so the warblers also benefit from the sentinel effect of the kingbirds (Smith et al. 2008). The two Hawaiian honeycreepers that in autumn descend every evening to lower elevations on Mauna Kea, in summer fly to roost at higher elevations more than 500 m above and several kilometers from their daytime feeding sites; of the circa 1750 mm of rain that falls annually on Mauna Kea, only 44 mm usually falls from June to August, and 80 percent of that is during the day. Roosting in a dense forest shelters the birds from what little rain may fall after dark, and they are at an elevation above the nightly fogbank (MacMillan and Carpenter 1980).

Large birds that cannot conceal themselves in wind- and rain-buffering vegetation seek other protective features. In northwestern Patagonia of Argentina, Andean Condors (*Vultur gryphus*) have traditional roost sites on cliff ledges that are often used communally. They favor cliffs where the lowest ledges are high off the ground, so that no predators can reach them. These ledges occur on cliffs facing in every direction, but condors consistently choose ones facing east or southeast, while the prevailing winds, which also sometimes carry rain, are from the northwest (Lambertucci and Ruggiero 2013).

Ease of Take Off

Soaring birds like the Andean Condors on cliffs in Argentina also seek a roost site that is easy to land on and to take off from. The additional benefit of their eastern-facing ledges is that these receive the morning sun, which warms both the birds and the air, creating thermals on which the condors can launch themselves (Lambertucci and Ruggiero 2013). Similarly, the sites of communal roosts of North American Turkey Vultures (*Cathartes aura*), such as one studied in Iowa that was on the east-facing slope of a hillside, may be chosen because this orientation and topography generate thermals earliest in the morning, so birds can get aloft sooner to begin their search for food (McVey et al. 2008).

For Rüppell's Griffon (*Gyps rueppellii*), the need for cliff roosts—in the absence of large trees over much of its range in the Sahel and savanna zones of Africa—restricts its daily foraging range, especially in winter when poor flying conditions make vultures late to depart and early to return to their roost, so they cannot travel very far. Perhaps as further evidence of the scarcity of good roost sites, some are used by hundreds of griffons. Since these vultures do not need to eat every day, some may spend most of the day at the roost, preening, stretching, and gazing about. Much farther south in Africa, Cape Griffons (*G. coprotheres*) also roost on cliffs, while the White-backed Vultures (*G. africanus*) roost exclusively in trees, preferring dead ones, which give an unimpeded view and an easier takeoff. Not needing cliffs perhaps explains this vulture's wider range over the continent, overlapping both of the griffons and extending beyond. Being smaller, it can use earlier, weaker thermals to take off from trees (Mundy et al., pp. 105, 80, 127).

Proximity to Food and Water

For birds whose food is patchy in location or irregular in time, as well as for birds with special roost site needs, the distance from the roost that they must—or can—travel to search for food influences their lifestyle. Some birds, like the vultures dependent on thermals, have only a few hours each day with suitable flight conditions, limiting how far they can travel and when their aerodynamic curfew requires that they return to their regular roost or compromise on another that may be less safe. Small nonterritorial birds that roost communally for safety or for warmth and then disperse to feed are likewise limited by travel distances and may be more flexible about their roost locale than their feeding site; over the course of a season, both may shift. Shorebirds on migration and in winter have curfews imposed by the tides that expose and then cover the flats where they feed and some of the places they roost. Because their flights between food and rest are powered by energy from the food they consume rather than from thermals or updrafts, they are limited by the amount of energy they have available for these commutes. Rich feeding sites too far from safe roosts will go unused—as may superior roosts too far from feeding areas.

Large Soaring Birds

Among vultures, the species restricted to scarce cliff roosts cannot search as far from their roost as the species that use trees, if these are more widely available. In the Etosha National Park in Namibia, the two commonest vultures are the Lappet-faced Vulture (*Torgos tracheliotos*) and the more numerous White-backed Vulture. Both are tree-roosters, so for daily search area both species have an advantage over any cliff-roosting vultures also present. Of the two, Lappet-faced Vultures roost in smaller groups that are more dispersed, so collectively they are able to cover a larger area. Since Lappet-faced Vultures also consume carcasses of smaller animals, which White-backs do not, their more widely spaced, smaller roosts enable them to find more of these as well. In addition, despite their larger size, their lower wing loading enables them to take off earlier in the day. White-backed Vultures in Etosha sometimes roost in groups of more than two hundred; they may find more carcasses near their roost, while the widely scattered Lappet-faceds discover more over all. At the large carcasses, where mammalian scavengers and potential predators also gather, however, the Lappet-faced Vultures wait for many White-backs to land and perhaps overwhelm any jackals, so even if the Lappet-faceds' more dispersed roosts give them the initial search advantage, they benefit from the presence of their less cautious competitor (Spiegel et al. 2013).

In North America, there are some ecological parallels and contrasts in the much smaller vulture community, consisting of only the Black Vulture (*Coragyps atratus*) and the Turkey Vulture, which are sometimes competitive at roost sites close to food. The larger Blacks feed only at large carcasses, while the Turkeys will consume smaller ones as well. Where feeding together, the Blacks are dominant. During spring and summer, when vultures in the Southeast roost singly or in small groups, Turkey Vultures at transmission towers give way to Blacks, shifting to trees; if no Blacks come to the towers, the Turkey Vultures remain there all night. In natural situations, the Black Vulture roosts are always near a recent or current food source, and these are usually used for only a few nights. Black Vultures feed most often early and late in the day, so they favor roosts in closest proximity. Where food is reliable and abundant year-round, such as at garbage dumps, the two vultures will roost together at traditional sites (Stewart 1978).

Among Bald Eagles, distance between roost site and feeding area is, at some locales, more defined by age class and foraging skills. In autumn at Glacier National Park, Montana, migrating eagles are attracted to creeks where salmon spawn. Hundreds may roost in conifers at four sites 3–7 km from a 4 km stretch of one creek where they were monitored. More subadult birds roosted closer to the creek and tended to leave their roost earlier in the morning to be the first to get to the dead salmon that washed up onto gravel banks the prior night. Adults, meanwhile, remained at the roost longer into the morning and then pursued live fish (Crenshaw and McClelland 1989).

Smaller Landbirds

For some communal roosting birds, the roost is the stable element in their daily cycle while the feeding areas to which they commute change over time. For others, the significant base of operations is the feeding site. Radio-tagged European Starlings in New Jersey were found to use a succession of roosts that were most convenient to wherever they were foraging most of the day, called the Diurnal Activity Center. These centers are fields or feed lots where there is a reliable supply of food that may be available for days or weeks. Starlings tend to roost within a few kilometers, but not always at the same roost when others are equidistant; some used as many as five different roosts while commuting to a single activity center. The roosts may be three to twelve kilometers away from the activity center. Often, the roost site is determined by the presence of additional but less significant secondary feeding opportunities between the roost and the diurnal center, where the starlings may stop en route at the beginning and end of the day. When an activity center is exhausted, the birds move on to search for other reliable sources and to settle at a new convenient roost. It is likely that this pattern matches the communal roosting behavior of many other birds that move from one reliable feeding site to another over weeks and months;

tracking individual birds is necessary to confirm this (Caccamise and Morrison 1986).

The birds that have the shortest distance between food and roost are the African oxpeckers, which feed by riding on large native animals and now on domestic animals such as donkeys, horses, and cattle, picking ticks and other parasites from their skin and fur. The Yellow-billed Oxpecker (*Buphagus africanus*) favors buffalo and giraffe, but it also rides on white rhinoceros, several ungulates, and hippopotamus. It roosts at night on its host, with up to nine birds on one animal; during the breeding season, it also roosts communally in trees together with various starlings. This could be because host animals may move far from the oxpecker's nest during the night. The Red-billed Oxpecker (*B. erythro-rhynchus*) more often roosts in palms, reedbeds, and holes in trees and rocks, but sometimes also on large ungulates (Fry and Keith 2000, pp. 664, 667).

In dry regions, proximity of a roost to water may be a higher priority, especially for birds that feed on seeds, which lack the moisture that other dry-country birds get from insects or fruit. In Australia, the Bare-eyed Corella (*Cacatua pastinator*) and Slender-billed Corella (*C. tenuirostris*) roost communally in trees. Both these small cockatoos depart before sunrise to drink at a local source; they then disperse in open areas to consume grass seeds on the ground. The Bare-eyed's habit of invariably roosting near water was well known to early travelers, who found waterholes by following their returning flights at dusk (Forshaw 1989, pp. 146, 148). In Baghdad, Collared Doves (*Streptopelia decaocto*) roosting in gardens first fly to the Tigris to drink, then return to preen, and only later go off to feed (Hutson 1956).

Shorebirds

Perhaps the birds with the most complex equations about roost site are shorebirds that feed in the intertidal zone. For them, the factors include predation risk; disturbance rate; the energetic cost of remaining thermoneutral where they may at various times be exposed to intense sun, strong winds, profound cold, or heavy rain; the energetic costs of flying to the roost from wherever they have been feeding; the distance and cost of moving to another roost if disturbed at the first choice; and the available space for a flock of various numbers. In addition, since the availability of both feeding areas and roosts are governed by the tides, no two days are the same. Places near feeding areas that are good for midday roosting (because they provide unobstructed views) may be underwater a few days later.

These several and variable factors sometimes lead to compromise choices. At Roebuck Bay in northwestern Australia, Great Knots and Red Knots roosting during daytime high tides use some of the few options near their feeding areas. Here they are disturbed by raptors and by people, each time flying off and then returning. The total energy expenditure from these alarm flights is greater than the cost of flying 25 km to another site with less disturbance, but the heat stress from such a long flight in the middle of the day may make this prohibitive at this subtropical latitude (Rogers et al. 2006b). On the Pacific coast of Baja California, Mexico, shorebirds of twenty-five species wintering at a lagoon, Guerrero Negro, roost on floating masses of dried seagrass (*Zostera marina*) that, during the highest tides, are the only dry surface near their feeding areas. At one locale, the nearest alternative is sand dunes almost 10 km away. Egrets, gulls, and terns also use the seagrass to roost. The shorebirds prefer the masses that are drifting through the marshland with the current, while the larger birds favor the accumulations that have stabilized near the edge of the desert (García-Walher et al. 2023).

Moreton Bay, on Australia's subtropical eastern coast, supports sixty thousand shorebirds of forty-two species, including thirty-two species of migrants from Alaska, Siberia, and New Zealand. For seven of the twelve species closely studied there, proximity to good feeding areas was the best predictor of the roost sites most used; their preferred maximal travel distance reflected the size of each species. For all, their rate of food intake is constrained by time, prey abundance, and slow digestive rate, so some exposed tide periods may not provide the energy necessary for long flights to a roost where birds can wait out the hours until the next feeding opportunity. Thus, to avoid energy deficits, these birds must find a roost site close by that also provides safety and thermal neutrality. Additional site features of value are a view of the ocean, which allows birds to track ebbing tides, and a place where they can stand in shallow water, either by the sea or at the edge of mangrove forests, to avoid heat stress by cooling their legs (Zharikov and Milton 2009).

Hudsonian Whimbrels (*Numenius hudsonicus*) on migration between northern South America and Canada roost during the night on edges or high ground in marshes and on offshore and barrier islands. A study that used satellite transmitters on tagged birds in Atlantic Canada, the southeastern U.S. coast, and the coasts of northern South America and the Gulf of Mexico found that the distances to night roosts were 3.1–42.2 km (mean 12.3), requiring 3.9–52.1 minutes (mean 15.2) of flight each way. The birds using offshore roosts had twice the commuting distance and energetic costs of those using onshore sites; they may seek such sites because these have fewer predators. The costs ranged from 1.5–18.8 percent of their pre-migratory budget, with 30 percent of roosts exceeding 10 percent. This matches the energy costs known for other species of migratory shorebirds in Europe. The whimbrels on northward migration in Louisiana had the greatest commuting costs, but they were going to and from rice fields, where they could feed all day, unlike the birds using tidal flats, and were therefore able to accommodate longer flights than the birds that had fewer hours to feed. In addition, the coastal feeders were concentrating on fiddler crabs (*Uca spp.*) that are digested slowly, thereby further limiting foraging time (Watts et al. 2021).

Fidelity to Roost Sites

Varying degrees of fidelity to roost sites reflect the ecology of each species over the course of its annual cycle and lifetime. The use patterns of nonmigratory, territorial species greatly differ from those of migratory and nomadic species. Among some species that take several years to reach maturity, adult and young birds have distinctive roosting habits. Solitary and flocking species differ, and among flocking species some may be social year-round or only seasonally. Some birds use a roost site for a single night or only a few before moving on, others remain at one or rotate among several for an entire season, while some birds continue for years if not a lifetime at the same place. Communal roosting species with a limited number of safe sites may use them for generations.

Birds that typically move over portions of a large home range each day, or search for widely scattered food sources that are consumed in one or a few days, have little reason to return to a roost that requires a long trip. The Northern Ground Hornbill (*Bucorvus abyssinicus*) may have a home range of 260 km² in African savanna. It is a reluctant flier, usually pursuing its prey of small animals on foot, walking about in pairs or small groups that include young birds. In the nonbreeding season, the party roosts in trees where it finds itself at the end of the day; the cost of returning to where the birds began that morning is prohibitive, whether on foot or by flight (Kemp 1995, p. 32, 93). Emus (*Dromaius novaehollandiae*) are nomadic during the nonbreeding season and roost wherever they are at dusk. If local food continues to be available for several days, they may return to roost for those nights in the same spot (Davies 2002, p. 224). On a different scale, Common Redpolls (*Acanthus flammea*) wander about the boreal zone of the Northern Hemisphere for most of the year after their short breeding season, searching for crops of birch seeds. A redpoll banded one winter in Michigan was recovered the following year in eastern Siberia, 10,200 km away (Troy 1983). During the long Arctic nights, they may feed during moonlight to maximize their intake. Before they settle somewhere to nest the following spring, they may have traveled tremendous distances, likely requiring a different roost every several nights if not more often.

Birds that move about within a smaller area may have several roosts they use over the course of a year. In Africa, the Sacred Ibis (*Threskiornis aethiopicus*) feeds in both dry land and all types of freshwater habitats. It responds quickly to rainfall or flooding, moving to new feeding sites immediately after water rises there. Flocks roost in trees near their current foraging area, shifting roosts frequently. In Witwatersrand, South Africa, only three of twenty-eight roosts were used year-round (Hancock 1992, p. 214). In the Delta Region of Mississippi, wintering

Double-crested Cormorants (*Phalacrocorax auritus*) fitted with radio-telemetry transmitters were found to move among fifteen roosts between January and April, with most birds using two or three and only one using four, but some traveling as far as 398 km among roosts (King 1996). At a Black Vulture roost in North Carolina, the nightly turnover of marked birds during a five-year study was 34 percent. Juveniles were most faithful, young adults ranged far, and adults were intermediate in their movement. When there were disputes over perches at the roost, the loser was less likely to return the following evening (Rabenold 1987). Another study, in Florida, focusing on two Black Vulture roosts over three years, found that some birds used at least five sites, which may be as many as 152 km apart (Stolen and Taylor 2003). Among small birds, like the several European finches that move about in flocks within a relatively small area throughout the nonbreeding season, the roosting pattern is similar to these larger birds that also have scattered food sources rather than consistent ones: roost sites are used regularly, but each night may host a different number and each bird may use several (Newton 1973, p. 123).

Birds with individual or paired home ranges that they use during an extended season or through their lifetime have various degrees of fidelity to roosts within their territory. In New Hampshire, White-breasted Nuthatches (*Sitta carolinensis*) often sleep in holes excavated by Downy Woodpeckers, which, on other nights, sometimes use those same holes themselves. When the nuthatch leaves the hole in the morning, it regularly removes the feces that accumulated there during the night, while the woodpecker defecates when it exits; both, thereby, keep the hole clean for future use. In spring, a regular winter roost site of the male nuthatch sometimes becomes the nest choice of the female (Kilham 1971). Carolina Chickadees (*Poecile carolinensis*) in Tennessee spend the winter in small flocks, with each bird having several roosts within the flock's home range. These are under loose bark, in hollow limbs, or cavities excavated by Downy Woodpeckers or by the chickadees themselves for a nest the prior season. Several birds might each singly use any one of the cavities on different nights. None of these, however, are later used for nesting. The chickadees do not clean their roost; this may cause them to move every night, and the accumulated scent of a winter roost may make chickadees create a new nest site that will not attract so many predators (Pitts 1976).

Larger, long-lived birds that maintain a year-round territory for the course of their adult lifetime are often faithful to specific roosts. Great Horned Owls (*Bubo virginianus*) seek dark, concealed spots to spend the day. These may be near wherever the bird was last hunting before dawn, but during the breeding season males typically have a favorite site not far from the nest used by their mate; often both nest and perch are used in successive years (Houston et al. 1998). Black Eagles (*Aquila verreauxii*) in Zimbabwe usually roost as pairs, in trees or on rocks near their nest. One pair sometimes used the same site, albeit not quite every night, for at least thirteen years (Gargett 1990, p. 57).

Knowledge of how long individual birds are faithful to a roost site, and how long roosts have been used by generations of birds, especially flocking species, depends substantially on how long the birds or the sites have been monitored. It is likely that long-lived sedentary birds like large raptors continue at the same sites as long as neither the bird nor the site itself is disturbed. Flocking species that roost communally in distinctive habitat features such as marshes, islands, or cliffs (chosen for safety or warmth) may well have used them for centuries. Manmade sites that provide the same benefits, such as the oil refineries now used by Purple Martins in Brazil and European Starlings in Great Britain, may in due course have similar histories. A 1934 review of European Starling roosts in Great Britain cited one on Lundy Island known since 1754 and another in a Devon reedbed since 1798. Of 247 roosts, 19 were known for 40 or more years (Marples 1934). Similarly, a 1938 survey of American Crow (*Corvus brachyrhynchos*) winter distribution in New York state found six roost areas—if not continually the same precise spot—known for 40 years and one known since before 1811 (Emlen 1938). Continued use, especially for sites visited only seasonally, like the chimneys where flocks of southbound Chimney Swifts (*Chaetura pelagica*) gather for a few weeks every autumn, depends on the survival and memory of birds that used them the year before.

Summary

Every day, every bird must choose where it will roost. Whether it spends its entire life in the same small territory or stops for a single day at a new spot while on migration, the factors a bird considers may include safety from predation, warmth and protection from rain, snow, or wind, and proximity to food and water. At different seasons, in different stages of the annual and life cycle, and for different sexes, the choices about site and proximity to others will vary.

Since birds are most vulnerable when they are asleep, safer sites may be those more concealed from the perception of predators or those least accessible even if conspicuous. Sometimes optimal sites are nowhere near where birds spend the day feeding, and they travel greater or lesser distances to a more secure place. For other birds, safety comes from roosting in flocks, where at any given moment at least some birds will be vigilant and the odds of any individual being picked off are reduced. Urban locations and industrial sites, if they have fewer predators, also attract birds that may otherwise avoid built landscapes. Some shorebirds with no safe options during high tides spend a few hours flying in flocks over the sea to avoid hawks and falcons, while some seabirds that rest on the water form dense flocks that may reduce their chance of predation by fish, sharks, and marine mammals attacking from below.

These safety concerns must also be weighed with thermal ones. The coldest hours of the day are when most birds sleep. Small birds are generally less resistant to cold than larger ones, so they often choose spots that can provide or retain some warmth, but even birds like eagles will commute to places with more protection from wind and rain, which affect birds' ability to maintain their body temperature. Outside the breeding season, few birds sleep on or in their nests, but some make or seek out other shelter, such as tree cavities or the insulated old nests of other birds. Others fly far from their daytime territory or to different elevations with better microclimates.

Proximity to food and water also affects the choice of roosting sites. Birds like scavengers and fruit eaters that feed on patchy food sources sometimes roost near it until the supply is exhausted, when they move on in search of more and shift to another roost. Birds with a reliable and proximate food supply can be more selective about the optimal roost in close range. Shorebirds dependent on safe high-tide roosts have the most complex equation in finding roost sites, because the places they may use for a few days may be underwater the next few. They must balance the distance they are able to fly in different weather conditions, hot and cold, with the safety of the site and whatever protection it may provide from the elements.

All these factors affect each bird's fidelity to roost sites. Permanent residents may have one or a few among which they rotate. Some long-lived birds likely use the same ones throughout their lives, as long as these are not disturbed. Other birds change the type of their roost site by season, while migratory birds must look for the best place they will use only once or during a few weeks or months. Birds that roost in flocks in sites with particular safety features like isolation or inaccessibility to predators may use the same places for generations or centuries.

4

Specialized Roost Sites

Some birds, especially small ones that sleep alone without the benefit of flock mates alert for danger, go to greater efforts to create or secure a roost site that can protect them from predation, weather, and other hazards through the hours they are most vulnerable. In environments of extreme cold or with a scarcity of safe sites, some larger birds also seek concealing shelter or fabricate it for themselves. Cavities, burrows, tunnels in snow, as well as specially constructed dormitories and roosting platforms are among the structures birds use to meet these needs. Still other birds go to the opposite extreme—they simply take to the sky, rising beyond the limit where anything would pursue them.

Cavities

Holes in trees, be they created deliberately for roosting, co-opted from some other excavator, or naturally occurring, are available in all wooded habitats. While scarce compared to the number and variety of places a small bird could secrete itself in foliage or among branches, and used by a relatively small portion of the birds in any wooded habitat, they are highly sought-after and consistently used by some species in all latitudes. In warm regions, the principal benefits may be concealment from predators and protection from rain and wind, while in colder latitudes and seasons it is the insulation that the cavities provide that is perhaps the most significant factor. Whether a cavity is occupied or empty, as

ambient temperature falls during the night, the cavity interior temperature drops at a lower rate.

Not all the cavities available to any bird are equal in benefit, due to their orientation, substrate, structure, and condition. Cavities in healthy living wood are warmer than those in dead or dying wood, especially during severe weather. Both the bark on living wood and the water content in cavity walls may provide some thermal benefits. Cavities are also warmer where the tree diameter is wider at the height of the cavity, because during the night the interior of a trunk remains considerably warmer than its surface. Thus, holes in larger trees are more valuable than in small ones, where a cavity is surrounded by thinner walls. The size of the cavity entrance also matters—small holes make for warmer interiors than large ones. And within the cavity, greater depth and the presence of any insulating material increase heat retention (Paclik and Weidinger 2007).

Woodpeckers, Creators of Their Own Shelter

Woodpeckers are the most widespread and accomplished excavators of tree cavities. The nests and roosts they make for themselves later provide many other birds, as well as other animals, homes used for a season or perhaps a lifetime. Without woodpeckers, these cavity dwellers would be limited to naturally occurring holes in trunks and branches that may be less structurally sound or appropriately shaped. For these secondary cavity users, the number and quality of roost sites created by woodpeckers directly impact their demography and distribution. For woodpeckers themselves, the simple availability of living and dead trees of various ages and dimensions all influence their choices when making cavities for various purposes. Dead wood is easier to excavate and may therefore, at higher latitudes, be used more often in spring and summer for nesting, which usually requires a larger cavity, while the cavities made for winter roosts may preferentially be in living wood. For woodpeckers that typically roost alone in the nonbreeding season, the effort required to make these smaller holes may be less than for the more capacious nests. In warm regions, dead trunks and branches may be preferred throughout the year.

Some woodpecker species create and use roost cavities more than others, and some make more distinction between roosting and nesting cavities. The use of cavities probably originated for nests, and the excavating skill was later applied to roosts for other seasons among woodpeckers that then sought another site. In most species, both sexes participate in digging the nest hole, but at night the male sleeps in the new hole when it is big enough, as well as when it contains eggs and nestlings, while the female roosts outside somewhere nearby (Walters et al. 2002). Males may continue to use the nest hole as a roost after the young have fledged. In many species, males dig more holes specifically for roosting, with females using old ones and sometimes later adopting one for a nest (Skutch 1985, p. 35).

Nonmigratory woodpeckers make roost holes at different seasons. Red-headed Woodpeckers (*Melanerpes erythrocephalus*) in Virginia excavate most of their cavities during early autumn, when they are also caching acorns to retrieve during the winter. They favor trees in which they can do both, usually oaks with dead branches (Nickley and Bulluck 2019). Downy Woodpeckers (*Picoides pubescens*) each make a few roosting cavities for themselves between September and January, peaking in November (Kilham 1983, p. 6). Having alternatives enables them to move easily if one hole is damaged or if some danger arises. Within their shared year-round territory, male Downies prefer higher sites than females; this niche partitioning may enable a pair to stay closer together during the nonbreeding season (Grubb 1982). In the larger North American species, Pileated Woodpeckers (*Dryocopus pileatus*) roosts in former nest holes, with males in those of better condition (Kilham 1983, p. 82), while the Ivory-billed Woodpeckers (*Campephilus principalis*) that James Tanner (1942, pp. 58–59) watched in Louisiana in 1937–39 dug new roost holes near their previous nest in April, just after their young fledged. One year, the male and female each used a hole in the same tree.

Migratory woodpeckers are likely more flexible about roost sites, especially when using them for only a night or two and where finding an existing unoccupied cavity in any unfamiliar stopover may be difficult and time consuming. A study in British Columbia attached geolocators to seventy-six Northern Flickers (*Colaptes auratus*) and found that over the course of a year the birds spent 63–90 percent of their nights roosting in cavities, including 49–89 percent while on migration. Throughout the nonbreeding season, most flickers spent no more than three nights in a row roosting outside a cavity, while during winter some roosted consistently in cavities and others only on alternate nights (Gow et al. 2015).

Some woodpeckers have quite specialized roost cavities. In the non-breeding season, the Grey-and-buff Woodpecker (*Hemicircus concretus*) of Borneo roosts in stable groups of three or four birds that each sleep individually in cavities that males excavate in dead branches or dead trees; one monitored group had seventeen cavities, another had thirty-four. The cavities are stacked or in rows only several centimeters apart, with as many as eighteen on a single branch, and each cavity is very shallow, 2 to 3.1 cm deep. Each bird in the group moves to a different cavity every night. The benefits of this social and design system include reduced risk of predation for any individual, especially since each bird changes its roost nightly. This also reduces the parasite accumulation in each cavity, while their shallowness may discourage cavity usurpers (Lammertink 2011).

Other Cavity Users

Among the three families most closely related to woodpeckers, the barbets and toucans also excavate cavities for nesting as well as specifically for roosting, while the parasitic honeyguides, which deposit eggs primarily in the nests of various other cavity nesters, roost in more open situations. Neither barbets nor toucans have bills shaped or strong enough to hammer into very hard wood; they dig into softer wood, use old woodpecker holes or natural cavities, or take over the nests of other birds. Many barbets roost in groups, sometimes all fitting into a single large hole, as once observed of sixteen Prong-billed Barbets (*Semnornis frantzii*), or using several adjacent holes, while the Pied Barbet (*Tricholaemia leucomelas*) roosts singly and sometimes uses the abandoned nests of swallows and weavers. Crimson-fronted Barbets (*Megalaima rubricapilla*) excavate their own cavities, specializing in the underside of small dead branches, sometimes later expanding the cavity for nesting purposes (Short and Horne 2001, pp. 60, 182, 279). Toucans monitor potential roost sites for availability, just as they check fruiting trees to see whether the fruit is ripe. They also use their long bills to reach into the nest cavities of other birds to prey on eggs and young inside, thereby accelerating the availability of the hole for their own use (Skutch 1958).

Many owls, especially smaller ones, roost and nest in cavities, natural or created by woodpeckers, and now in specially designed birdboxes. Their brown striped and barred plumage so well matches the bark and grooves of trees that an owl resting at the entrance of its cavity during the day, as they often do when the sun reaches it, is less noticeable to potential predators than is the dark circle of an unoccupied entrance. Finding the best cavities may be a learned skill. A German project that followed twenty-four adult and seventy-five juvenile radio-tagged Little Owls (*Athene noctua*) from July through November found that they preferred sheltered roosts, especially tree cavities with more than one entrance. They used increasingly sheltered roosts as the weather grew colder, with more than 80 percent of adults and 50 percent of juveniles in these during freezing weather. The juveniles used less sheltered sites while they were dispersing—perhaps because they had less skill finding good ones or because these were already occupied by resident adults—than where they settled for the winter. Even then, however, their roosts were not as effective against predation or cold as those of adults; juveniles were also more predated, most likely by diurnal raptors, than adults (Bock et al. 2013).

Nuthatches and tits are the best studied of the several passerine families that frequently use cavities. White-breasted Nuthatches (*Sitta carolinensis*) and Downy Woodpeckers sometimes alternate in the same roosting hole, likely made by the woodpecker, both fastidiously keeping it clean. The nuthatch prefers cavity entrances two to three times its body size. This may enable it to slip out the entrance to escape the paw of a mammalian predator reaching in; the woodpecker's body could fill the entrance while it uses its strong bill to peck at the intruder (Kilham 1971). White-breasted Nuthatches normally roost alone, but in extreme cold several may gather together; one roost once held twenty-nine birds. The much smaller Pygmy Nuthatch (*S. pygmaea*) cannot retain its normal body temperature in cavities. Even in summer, it routinely uses a roosting cavity larger than its nest, accommodating more than just the small family unit with its dominant pair, young, and occasional helper. These cavities may have a large entrance as well as cracks and additional openings that provide some ventilation. In autumn and winter, when temperatures may drop to −40°C, they choose cavities with narrower entrances and thicker walls but more interior capacity. More than one hundred have been counted in a single cavity. The birds will be several layers deep, and those at the bottom risk suffocation; this risk is greatest during periods of heavy snowfall, when the birds may not leave the hole at all for more than forty hours (Harrap 2008, pp. 114, 116).

The fifty-six species of tits and chickadees all nest in cavities—natural, woodpecker-made, and manmade—and many of them use these cavities for roosting in other seasons, depending on weather and availability. In the Netherlands, Great Tits (*Parus major*) consistently sleep in cavities from November to at least March, always singly. The cavity used by the female in March or April becomes the nest cavity, but it is never used by the young after they fledge. At that time, the female, following the male, shifts to sheltered spots against a tree trunk or in foliage, with the return to cavities beginning in August (Kluijver 1950). Black-capped Chickadees (*Poecile atricapillus*) and most other tits similarly roost singly, choosing roosting cavities that are smaller than nest cavities, likely because the more compact the space, the more easily the bird can generate enough heat to keep it warm, even in those species that regularly enter hypothermia on winter nights. Where cavities are scarce, dominant individuals in the small bands that forage together during winter days exclude the subordinate ones from the better sites (Smith 1991, p. 243). Among the few tits known to roost communally are the Siberian Azure Tit (*Cyanistes cyanus*), which is not known to use hypothermia, and the White-bellied Tit (*Parus albiventris*) of tropical African forests and woodlands at elevations of 1000–3400 m, which may be cool enough to justify sharing a cavity (Gosler and Clement 2007, p. 686).

In other families, cavity-use outside the breeding season depends on the birds' previous experience and the immediate situational need. Among the swallows, only the species that nest in holes sometimes also roost in them. Wintering Tree Swallows (*Tachycineta bicolor*) normally roost in large flocks in trees or wetlands, but in cold weather some will resort to cavities, where as many may gather as will fit. Here, they often also enter torpor and remain until conditions warm up, some dying in the roost, from suffocation or starvation, since they were likely in poor condition when they sought out a cavity (Robertson et al. 1992).

The enormous flocks of the cavity-nesting European Starling (*Sturnus vulgaris*), well known for the aerial gyrations they perform before descending to their communal roosts, are, depending on the month, a mix of nonbreeding, migrant, local, and young birds. Some local birds in the area, however, remain on their territories throughout the year, with pairs singly or together roosting in the hole they nested in and will again the following spring; continual occupancy ensures that no one else takes it (Feare 1984, p. 65). Other starling species, such as the African Superb Starling (*Lamprotornis superbus*), which sometimes builds a nest of twigs and other times uses holes in trees or cliffs, may resort to cavities of a different sort for roosting, the vacant nests of various weavers (Feare and Craig 1999, p. 219).

Burrows

In habitats where trees are scarce or absent, some birds roost in burrows in the ground. Burrows may be used consistently or selectively. Some capable excavators like woodpeckers use both soil and trees, for different purposes. In South American regions with few trees, Chilean Flickers (*Colaptes pitius*) and Campo Flickers (*C. campestris*) dig roosts in banks and in termite nests on the ground. They also feed on the termites. For nesting, however, they prefer trees (Skutch 1985, p. 35). Some entire families of birds, including the kingfishers, motmots, and bee-eaters, that live in forests and at the edge of woodlands also nest and roost in burrows, even where tree alternatives are available. A few parrots and passerines dig burrows in termite nests, both those in trees and on the ground, where these are a more available or more workable alternative to other substrates.

Other birds use burrows exclusively for nesting. Seabirds such as shearwaters, petrels, storm petrels, puffins, and some of the small penguins all spend the rest of the year at sea and are not known otherwise to occupy burrows. Similarly, highly migratory Bank Swallows (*Riparia riparia*) do not use burrows except during the nesting season; then, in cold weather, a few may gather in burrows that do not contain their nests, but like most other swallows, they otherwise roost in trees or wetland vegetation.

Benefits and Hazards

Burrows in soil offer benefits similar to tree cavities. They are insulated from cold, wind, and rain, and they can be oriented to the afternoon sun to capture heat at the end of the day. Just as woodpeckers may orient their nests and roosting cavities for warmth or a drying breeze, burrow excavators also consider thermal conditions. The burrows of European Bee-eaters (*Merops apiaster*) tend to face south or southwest. At one colony in southern Spain, the soil surface temperature above these burrows, when occupied, ranged from 13°C to 51°C, while within the nest chamber it was a steady 25°C (Fry 1984, p. 233). The Tibetan Ground Tit (*Pseudopodoces humilis*) lives in alpine meadows at 2500–5000 m; it excavates a nest burrow in soil in the spring and later another one for roosting the rest of the year. This winter burrow is oriented to the afternoon sun and away from prevailing winds. The burrows are 100–160 cm long, with chambers 20–40 cm deep to avoid colder surface soil temperatures. Longer tunnels reduce wind penetration, while shorter ones may be warmed sooner (Ke and Lu 2009).

Burrows, however, come with some hazards not found in tree cavities. They are accessible to ground predators, especially snakes, which can slip in more easily than most larger mammalian predators. Horizontal burrows excavated into the banks of streams or rivers have the risk of flooding if the waters rise, while entire banks sometimes collapse after heavy storms. In South Africa, the Ground Woodpecker (*Geocolaptes olivaceus*), the ecological counterpart of Western Hemisphere flickers, excavates a burrow one meter long in vertical banks; the burrow is always slanted upward (Skutch 1985, p. 35).

The birds that excavate into more or less flat ground rather than into banks have additional hazards. African White-throated Bee-eaters (*Merops albicollis*) nesting in dunes at Lake Chad sometimes have their burrows obliterated by sandstorms and trampled by elephants and buffalo

(Fry 1984, p. 231). Vertical burrows in open areas may accumulate rain if the roosting or nesting place at the burrow's end is not dug at an upward angle. D'Arnaud's Barbet (*Trachyphonus darnaudii*) in East Africa digs a hole straight into the ground, as much as 90 cm deep, but with the nest chamber off to one side of the main tunnel, well above the bottom, where rain water may accumulate. Adults roost in the nest and fledglings return to it in cold, windy, or wet weather, indicating that the design is effective (Short and Horne 2001, p. 130–31).

Another significant cost is the accumulation of parasites, especially since birds using burrows are less likely to move as frequently as many cavity roosters, because burrows, either made or expropriated, are scarce. Burrows made by birds themselves require substantial time and energy, so few birds are likely to have many available. And, after making the effort to create a burrow for roosting, some birds use it as a nest, which substantially increases the parasite load. Sedentary species tend to make new burrows for themselves after the nesting season, while migratory birds that have left the burrow empty for many months may be able to use it again. Nonbreeding African Pied Kingfishers (*Ceryle rudis*) usually roost in trees, but sometimes adults and immatures will together use unoccupied nest burrows, perhaps not going all the way to the end where the old nest remains (Woodall 2001).

Birds That Dig Their Own Burrows

While woodpeckers may excavate when alone, other burrowing birds such as motmots, jacamars, and kingfishers usually work in pairs no matter what the site. One perches outside the burrow to watch for danger that the bird with its head inside the hole could not see. Typically, birds loosen and break off the substrate with their bills and then kick it behind them (Skutch 1985, p. 37).

Migratory birds with a brief breeding season must make a new burrow when they arrive or use an old one still in good condition. Sedentary birds have the advantage of being able to create a new burrow whenever the soil conditions make work easiest. In the highlands of Guatemala, Blue-throated Motmots (*Aspatha gularis*) begin breeding in early April, but they dig the burrows they will use months ahead, soon after their young have fledged from the previous burrow. It is easiest to excavate in June and July, when the rains have let up and the soil in washouts and roadside banks is neither muddy nor dry and powdery. The motmots then work as a pair to dig the burrow, and they roost in it together every night from completion until eggs are laid the following year (Skutch 1989a, pp. 100–1). Similarly, some African bee-eaters roost in their burrows year-round, with the Red-throated Bee-eater (*Merops bullocki*) excavating its burrows at the end of the rainy season, three to four months before nesting begins (Fry 1984, p. 230). This long-term occupation ensures that burrows will not be usurped by a neighbor or any other animal, and it will give the owners time for any repairs or expansion, should these be necessary.

Flightless New Zealand kiwis are limited to the ground for roosting and nesting. The nocturnal Great Spotted Kiwi (*Apteryx haastii*) is found only in a small portion of the South Island, where it lives in wet mountain forests up to 1200 m. Within its territory, a kiwi may have one hundred roosting dens, either burrows excavated by the bird or natural holes in the ground; each bird uses a different den every day, with about 40 percent of pairs using the same one on the same day. The dens are at least one meter wide and 50 cm high, with two or more small entrances. Into each den, the kiwi brings vegetation to make a mat upon which it rests. Since New Zealand had no native ground predators, neither snakes nor mammals, these burrows, and the ones later used for nests, must be for warmth and dryness in the damp forests; they may today provide some protection against the introduced ground predators now widespread in New Zealand (Davies 2002, p. 256).

Of the more than two hundred species of owls, only the Burrowing Owl (*Athene cunicularia*) has become so terrestrial that it lives in burrows. Found from western Canada to southernmost Argentina, it inhabits prairies and extensive grasslands that nearly everywhere are also the home of various ground-dwelling mammals—including prairie-dogs (*Cynomys*), viscachas (*Lagostomus*), and badgers (*Taxidea*)—that provide a network of burrows. Only in Florida and the Caribbean islands, where there are no open-country burrowers, must this owl dig its own. There,

Burrowing Owls dig with their beak and then kick the soil backward with their feet. They can complete a three-meter burrow in two days. Both roosting and sleeping take place at the mouth of the burrow, which may prevent predators from entering (Haug et al. 1993).

While most parrots roost in trees, a few excavate burrows. In South America, Patagonian Conures (*Cyanoliseus patagonus*) breed in colonies, with each pair digging a zigzag burrow up to three meters long into a sandstone or limestone cliff or in an exposed sandbank. The endangered Lear's Macaw (*Anodorhynchus leari*) of the arid caatinga of northeastern Brazil roosts high on cliff faces in burrows created by weathering. While the entrance to each burrow is only large enough for a single bird to enter at a time, as many as four have been found in some burrows (Forshaw 1989, pp. 471, 392). Burrows in termite nests, in trees and on the ground, are used as nests and roosts by a few parrots as well as by some woodpeckers, kingfishers, jacamars, and puffbirds. Arboreal termitaries are favored by species that live in the forest canopy and never come near the ground. In New Guinea and some adjacent islands, two pygmy parrots (*Micropsitta*) excavate their nests and roosts in large termite nests high in trees. As many as eight have been found together in a termitarium burrow that was not being used as a nest (Forshaw 1989, pp. 162, 164).

Burrows Later Used by Other Birds

As with tree cavities, burrows may be used by other species that cannot excavate them themselves. High in the Andes, well above the tree line, Andean Flickers (*Colaptes rupicola*) are found up to at least 5150 m; they excavate tunnels in soil that are later used by many smaller birds for nesting and roosting (Hardy et al. 2018). In the alpine zone of Mount Kenya, adult Red-tufted Sunbirds (*Nectarinia johnstoni*) have been seen roosting in the deep holes made by Mountain Chats (*Pinarochroa sordida*) in matted dead-leaf clusters of the tree groundsel (*Senecio keniodendron*), with as many as three adults entering one hole, while immature birds roost in unused nests (Williams 1959).

Birds will also use the abandoned burrows of other animals. In southern Africa, the Southern Anteater-chat (*Myrmecochichla formicivora*) digs farther into the burrows made by aardvarks, hyenas, porcupines, springhares, yellow mongoose, and ground squirrels to make a cavity suitable for its own much smaller size. The anteater-chat is a cooperative-breeder; its burrows may hold as many as five individuals, and the birds of each territory may use up to ten burrows over the course of a year (Collar 2005, pp. 547, 725).

Snow

Snow, like soil, is an insulating material useful for making a burrow or cavity. In some places snow may be the only material available when it covers the ground, already likely frozen and impenetrable, and where trees, if present, do not provide much shelter. Snow is the regular roosting place for many species of grouse through the long winters in high latitudes, as well as for some very small birds like tits and finches that share these latitudes. Even much farther south, when snow puts a thick layer over the ground and the weather conditions are extreme, small birds make it their shelter. If the snow is soft, birds generally fly or dive into it, thereby leaving no trail of footprints for a predator to follow. When the snow is harder, birds must dig into it. The risk is that moist snow or rain may later freeze, trapping a bird in its burrow.

Grouse

Even for relatively large birds like grouse, the thermal benefits of roosting in snow are significant. In northern Finland at 65°N, where the winter temperature may fall below −40°C, the air temperature surrounding Capercaillie (*Tetrao urogallus*) in their snow burrows was measured at 11°C, while the snow itself varied between −7.5 and −11.5°C and the outside temperature was usually around −10°C. Individuals kept in an aviary above the snow through the same nights had considerably higher body temperatures, indicating that they needed to consume more energy to maintain warmth, while those in burrows were less alert, a sign that they were also using hypothermia to conserve energy. In other experiments with Hazel Grouse (*Tetrastes bonasia*) in Siberia, when the

Burrows in snow insulate Black Grouse from low temperatures and conceal them from predators through the long winter night and during most of the short day.

air temperature was below −40°C, the burrow temperature could rise to near 0°C (Marjakangas et al. 1984).

Reduced risk of predation is another important benefit of snow burrows. Black Grouse (*Tetrao tetrix*) in Finnish forests are conspicuous on winter days during the few hours they leave their burrows to feed in trees. If they roosted there, they would be vulnerable to Northern Goshawks (*Accipiter gentilis*) and Eagle Owls (*Bubo bubo*). Even in relatively mild winter weather, when the ambient temperature rises slightly above freezing and the temperature in a burrow would be lower, the grouse usually roost in snow. They also roost in snow when wet conditions increase the risk of the surface freezing, making it hard for grouse to emerge in the morning. Over 137 days of one study, no grouse that roosted in snow were found to have been predated, while there were remains of birds taken by goshawks during the hours the grouse were feeding. In contrast, Swedish Willow Ptarmigan (*Lagopus lagopus*), in their camouflaging white winter plumage, do not usually roost in burrows except when the minimum temperature is below −10°C (Marjakangas 1990).

Much farther south, Greater Sage-Grouse (*Centrocercus urophasianus*) in Nevada make several different uses of snow, depending on local conditions and availability. They sometimes roost in a shallow depression or open bowl in snow, with their heads and backs exposed, and other times they immerse themselves totally in a burrow. These may be in soft drifts on the lee side of shrubs. The birds tunnel into one side of the drift and later emerge on the other side; the entrance hole is plugged when the snow collapses behind the bird. In more open, level areas, burrows are longer and deeper, but they are only made when soft, dry snow exceeds 25 cm (Back et al. 1987).

Grouse are able to spend most of the winter roosting in snow with only brief periods for feeding because they store food in their crop that they can digest when back in their burrow. Willow Ptarmigan at 73°N in Alaska ingest enough to carry them through the following day, a benefit if that day is too stormy to emerge and feed (Skokkan 1992). The Black Grouse studied in Finland spent as much as 94 percent of the twenty-four-hour cycle in their burrow, emerging to feed once a day, sometimes for only forty-two minutes. The time spent in the burrow did not correlate with the extent of daylight as this increased over the winter, but instead with the ambient temperature. It may be that the energy required to generate heat in the snow burrow is more than a Black Grouse can do twice daily at 65°N. Farther south, in Sweden at 61°N and in Switzerland at 46°N, by March, Black Grouse come out of their burrows to feed in both morning and afternoon; in Switzerland, they were active for 1.5–2.5 hours in the morning and 1–2 hours in the afternoon. In contrast, Capercaillie are more active during winter afternoons, perhaps avoiding the higher energy costs of the colder mornings (Marjakangas 1992).

If conditions such as icy snow prevent grouse from retreating to burrows, they must find an alternative site. Black Grouse use conifers or rest on top of the snow (Marjakangas 1992). In south-central Sweden, where a maritime climate may make snow available in some winters for only a few weeks, Hazel Grouse select Norway spruce, which are shorter and denser than canopy trees, and roost low in them, seeking the places with greatest thermal benefit, from wind if not from cold (Swenson and Olsson 1991). But snow is clearly preferred when available. In winter when there is no snow, Ruffed Grouse (*Bonasa umbellus*) roost in trees or on the ground, but if snow begins to fall while a grouse is roosting on the ground, it stays in place and lets the snow cover it (Rusch et al. 2000).

In Missouri, at the southern edge of the range of Ruffed Grouse, twenty birds fitted with radio-telemetry transmitters were found never to use the same roost site more than once. Snow was only intermittently deep enough for roosting, at more than 20 cm; five of the six birds tracked at the time were using it. For most of the winter when no or too little snow was available, the preferred site was the canopy of cedar trees, which were denser than all the available deciduous trees and had the lowest wind speed. Of the different sites, snow provided the best insulation—the burrows were a consistent 1.5°C while the air temperature varied from −10° to −31°C. The burrows enabled the birds to reduce their metabolic heat production more than any other site (Thompson and Fritzell 1988).

Small Birds

A few small birds that normally roost on the ground, where they scrape a shallow depression for themselves, do the same in snow. The Black Lark (*Melanocorypha yeltoniensis*), wintering primarily in Russia and Central Asia, makes depressions in snow 10–13 cm wide and 5–10 cm deep, usually reaching the ground surface. Its winter feeding method is also to scrape away snow to reach seeds on the ground below, so roosting in snow is not as substantial a behavioral adaptation as for other small birds that create burrows for themselves. Eurasian Skylarks

(*Alauda arvensis*) and Horned Larks (*Eremophila alpestris*) are also known to roost in snow as they would on the ground (de Juana et al. 2004, pp. 515–17).

Among the small birds that actually burrow into snow are Goldcrests (*Regulus regulus*) and several tits, finches, and sparrows. The Goldcrest, the smallest bird wintering in the Eurasian boreal forests, with an average weight of 6 g, spends the entire winter day foraging and may gain as much as 1.5 g. This generates the energy required to maintain a steady body temperature without resorting to hypothermia. For a bird this small, however, the energy saved from hypothermia is not enough to compensate for the energy required at dawn to raise its body temperature to its normal daytime level (Reinertsen et al. 1988). Its usual strategy is to huddle together in small groups, most often in trees, but in Finland, Goldcrests have also been found in burrows that go through snow and that then terminate in a cavity in the soil beneath it. One group of three was seen entering such a burrow so directly that it was clear they had used it on previous nights (Lagerström 1979).

Larger than Goldcrests, Siberian Tits (*Poecile cinctus*) and Willow Tits (*P. montanus*) in Siberia also use snow burrows. A Willow Tit can dig a tunnel 20 cm long in ten to fifteen seconds, using its head, wings, and feet. They use the same burrow for extended periods, evidently able to locate it even after more snow has fallen (Smith 1991, p. 243). In Siberia, several tits and the Common Bullfinch (*Pyrrhula pyrrhula*) have been found roosting in the nests of small rodents, reached through snow. How they located these is unknown (Novikov 1972).

In Finland, Common Redpolls (*Acanthis flammea*) feed primarily on birch catkins during the winter and roost close by in snow near the trees where they have recently fed. Some drop straight from the tree and may burrow within one meter of others, then they dig a tunnel as many as 65 cm long some 6–11 cm below the surface. Within the tunnel, the bird remains in a slightly enlarged area, and it departs through a different place, sometimes leaving an impression of its wings in the snow. If the presence of droppings is a good indication of how long a redpoll has spent in a burrow, some are occupied for the night, while burrows without droppings were likely used only for briefer periods between feeding bouts. The burrows used only briefly are not as deep as the nighttime ones, but they are equally as long. When the ambient temperature is −20° to −40°C, the temperature 10 cm below in the snow is already 15–20°C higher before a redpoll has entered. The absence of wind further reduces heat loss within the burrow (Sulkava 1969). In the Adirondacks of New York state, a flock of 150 to 200 redpolls was observed during a cold and windy February day, with about half the birds at any one time excavating small caves in the snow on a garage roof; the birds used their beak and feet to move the snow to a depth of several centimeters, and then they turned around to face out. From there, redpolls flew to a nearby feeder, and the chamber they had created was quickly filled by another bird; others displaced a bird from its hole (Furness and Peterson 1987).

The benefits of snow roosting, compared to other potential available sites, can overcome some birds' usual social patterns. The Song Sparrow (*Melospiza melodia*) normally roosts singly on or near the ground in thick vegetation, but in early April 1977 some migrants arriving on Prince Edward Island, Canada, found the landscape covered in snow. A group of fifteen was discovered emerging in rapid succession from a hole in a snowbank; the hole was approximately 4 cm at its entrance and had a tunnel of 10 cm to a cavity 18–20 cm long by 11–14 cm wide and 8 cm high. Cold and a lack of options evidently motivated these otherwise solitary sparrows to excavate a burrow and huddle together for warmth (McNicholl 1979).

Caves, Glaciers, and Mines

In various tropical zones, caves are both the breeding and roost site for several species of swifts and for the South American Oilbird (*Steatornis caripensis*). For some other birds, caves in warm regions may serve only as roosts. In East Africa, Slender-billed Red-winged Starlings (*Onychognathus tenuirostris*) both roost and nest in caves behind waterfalls, but some caves are used only for roosting (Feare and Craig 1999, p. 235). In Amazonian Peru, a flock of White-eyed Conures (*Aratinga leucophthalma*), which sometimes roost in canebrakes on river sandbars, was once found within a cave, as far as twenty-five meters from its entrance. More than one hundred were there at midday, when the cave must have been cooler than the surrounding forest. That evening, after the birds had returned from foraging, the cave was occupied by 220, using holes and ledges. This species nests in tree cavities, so the cave was not also serving as a colonial breeding site (Forshaw 1989, p. 435).

Elsewhere, caves are sought out for their protection from rain, wind, and cold. On the sub-Antarctic islands inhabited by the two sheathbill species (*Chionidae*), they nest within penguin colonies, where they prey on eggs and chicks and scavenge carcasses, but in the nonbreeding season some roost in caves (Burger 1996, p. 548). Caves are also used by many birds living at high elevations in the Andes, some above 4500 m. Most high Andean birds nest in the wet season, a time of hail, snow, and cold rain. In addition to hummingbirds and ground-tyrant flycatchers that both nest and roost in caves, other birds more or less regularly roost in them. In Peru, Bolivian Goose (*Chloephaga melanoptera*), American Kestrel (*Falco sparverius*), Great Horned Owl (*Bubo virginianus*), Gray-hooded Finch (*Phrygilus gayi*), and additional ground-tyrant species have been found roosting in caves at 3400 m. None of these had a low body temperature, indicating that the thermal protection provided by the cave was as much as these birds needed (Pearson 1953). In Ecuador at 4115 m elevation, a small cave three meters high and six meters wide at the entrance and four and a half meters deep had roosting Ecuadorian Hillstars (*Oreotrochilus chimborazo*) and Stout-billed Cinclodes (*Cinclodes excelsior*); at 9 p.m. the temperature in the cave was 5°C, only a half degree warmer than outside, and at 6 a.m. the following morning it was 3.5°C (French and Hodges 1959).

Still higher in the Andes, between 5000–5200 m in the Quelccaya Ice Cap, where there are no caves, the crevasses and caves within glaciers provide similar shelter from snow, which falls throughout the year but especially during the wet season, when temperatures are −3.3°C and the wind speed averages 3.8 m per second. A few birds build their nests in glacial crevices—the White-winged Diuca-Finch (*Idiopsar speculifera*) is not known to nest anywhere else—and two species of seedsnipe and a sierra-finch have been seen at the glacier or flying toward it at dusk, presumably to roost there (Hardy et al. 2018). In Bolivia, more than two hundred Diuca-Finches have been found massed together at night in a glacier crevice at 5300 m (Skutch 1961).

Where buildings and other structures that provide similar thermal benefits are available, some cave roosters resort to them. The three North American rosy-finches descend in winter to somewhat lower elevations from the highest peaks in Alaska and the Rocky Mountains. Then, flocks routinely roost in caves, accessible buildings, shallow wells, mine shafts, and the nests of Cliff Swallows (*Petrochelidon pyrrhonota*) on rock faces. In Montana, a flock of sixty to seventy Gray-crowned Rosy-Finches (*Leucosticte tephrocotis*) was observed for several weeks from February to March roosting in a mineshaft with an entrance of 2 by 2 m. The birds perched at least 8–10 m below the shaft entrance. The air in the shaft was a constant 9–10°C while outside the overnight temperature sometimes dropped to −28.6°C. On some days, the birds remained in the mineshaft until 8 a.m. and returned at 3 p.m., spending as much as seventeen hours at their roost; the energy they saved by roosting in such a sheltered spot may have reduced the time they needed to feed to carry them through to the next day (Hendricks 1981).

Nests

During the breeding season, and sometimes beyond, nests provide many birds with protection from weather and concealment from predators that other birds must seek elsewhere. Some birds begin sleeping at the nest they are constructing even before it is complete, while others do not occupy it until it holds a full clutch and incubation must begin. Most of the birds that take advantage of their new nest as a roost are ones that select intrinsically sheltered spots. In addition, these sites justify occupation for as much time as possible to ensure that they will not be usurped by any other bird. Few birds, however, continue roosting in the nest after its function of supporting the young is complete.

Female Eastern Phoebes (*Sayornis phoebe*) begin roosting in their nest as soon as it is well formed, which may be three weeks before egg laying. This delay is likely due to the birds waiting for warmer weather and the emergence of more flying insects to feed their young; during this period of colder weather, the nest provides its occupant some insulation. Since phoebes usually place their nest on a ledge with a sheltering overhang, the female is not conspicuous when sitting on it. The scarcity of good sites may be an impetus to building the nest well in advance of breeding, before another phoebe can claim the territory. Phoebes monitored in Indiana were found to renovate nests from prior years; these were no longer cup-shaped, so the female at first roosted perched on the rim. As soon as the cup has been rebuilt and has some lining, females roosted in it. Once the first egg was laid, females roosted on the nest every night. Later in the spring, when a second clutch was laid during warmer weather, only 56 percent of female phoebes roosted on the nest before the first egg was laid. Meanwhile, males are presumed to roost elsewhere within the territory (Weeks 1994).

Roosting in the nest before incubation may in fact be common among flycatchers and not uniquely linked to thermal benefits or securing scarce sites. In the Yellow-billed Flycatcher (*Tolmomyias sulphurescens*), widespread in the lowlands from Mexico to Argentina, the female builds a suspended pouch with a tunnel entrance at the bottom, and she begins sleeping in it as many as ten days before laying the first egg. After the young fledge, she may continue using it alone for another four months, if rains have not destroyed it or other birds or wasps have not begun to use it while it is unoccupied during the day. The rest of the year, females, like males, roost in foliage. Other tropical flycatchers begin sleeping in the nest a few days before the first egg is laid, or before the clutch is complete (Skutch 1997, pp. 38, 89).

Among tits, which compete among themselves and with other small hole–nesters for cavities, females roost in the nest early. Female Black-capped Chickadees begin sleeping in the nest while it is still under construction, with the male roosting somewhere close by, likely also in a cavity. She continues through incubation and until the nestlings are at least twelve days old, capable of thermoregulation, and no longer need brooding. Then she finds another place to roost near the nest (Smith 1991, pp. 117, 131). In Scotland, Great Tits and Blue Tits (*Cyanistes caeruleus*) begin roosting in the nest cavity as soon as the first egg is laid, but without sitting so tightly that embryo development begins. Benefits to the female include reduced energy expenditure (because the nest hole is warmer than outside), reduced risk of predation, prevention of another tit laying an egg in the nest, theft of nest material, or takeover by other birds or animals (Pendlebury and Bryant 2005). The female Great Tit stays on the nest at night until the young are at least fifteen days old, the Blue Tit until they are at least ten days old (Perrins 1979, p. 162).

For all birds, the parent with the night incubation shift is roosting on the nest. In species where parents share incubation, that bird is usually relieved at dawn. If only one parent incubates, it usually leaves the nest, at least briefly, at about the same time relative to morning daylight, when it would depart its roost site at other seasons. Among seabirds that forage far from their colony, however, the incubating bird may be on the nest for several days while its mate is away; during these fasts, it may sleep during the day as well as at night, especially in cold environments where the lower body temperature that comes with sleep saves energy. Sooty Terns (*Onychoprion fuscatus*) on Ascension Island trade places at the nest every five or so days, when the foraging bird returns from a period over the sea during which it may never have landed. The Great Frigatebird (*Fregata minor*), which also stays aloft when on a foraging trip, returns at fifteen-day intervals. Some albatrosses may incubate for three weeks, perhaps sleeping much of the time, before being relieved (Skutch 1989a, pp. 10, 11, 12).

Among landbirds that share incubation and/or feeding the nestlings, the off-duty parent typically roosts relatively close by, if a suitable site is available. Colonial seabirds like gulls and terns that nest in the open often are next to their incubating or brooding mate through the night. Many swallow pairs spend the night together on the nest, at least while the young are small. In Chimney Swifts (*Chaetura pelagica*), both parents sleep on or near the nest; after fledging, the entire family continues using its place in the chimney for two weeks. Some species of "edible-nest" swiftlets (*Aerodramus*) continue roosting in pairs in or near their nest throughout the year; unlike the roosting habits of migratory birds, this may be a form of defense, if the nest can be reused in the following breeding season (Skutch 1989a, pp. 76, 75, 71).

Special Structures

Like the Verdins (*Auriparus flaviceps*) that build nests specifically for roosting at different seasons, each with its customized extent of insulating material, a small number of other birds build structures used only for sleeping. These may in form resemble the nests in which they raise their young, as does the Verdin's, or may be entirely different. The purposes also vary: some are built to keep the sleeping bird above water, others for warmth or concealment from predators. Whether or not these resemble the brood nest, it seems likely the constructing habit originated in nest building.

The transition is visible in the habits of American Coots (*Fulica americana*), which build several nests in addition to the one in which eggs are laid. These extras are used as roosts during the breeding season, as is the display platform, not as substantial as a nest, that male coots use at the beginning of the season. Other rails also build "dummy nests" and use these for brooding their chicks or for sleep, and some of the rails that live in forest far from water build sleeping platforms in trees (Skutch 1989a, pp. 16–17). In the mountains of New Guinea, Forbes's Forest-Rail (*Rallina forbesi*) leaves the ground to construct a domed platform in trees; as many as seven birds may roost in it (Taylor 1996, p. 121).

The Australian Magpie Goose (*Anseranas semipalmata*) breeds in extensive wetlands with no solid ground and water usually 60–90 cm deep. It has similarly expanded the nest-building habit to serve additional purposes that need a dry place elevated above the water. At the beginning of the breeding season, males build "stages" on which they court, preen, and rest, pulling down tall grass-like marsh vegetation to create a platform a few centimeters above the water. Later, when young hatch from the actual nest and the family moves through the marsh, adults regularly build similar platforms for themselves and their young. Even the goslings at only two to three days old go through the movements of building. For the ten weeks until the young can fly and the geese leave the marsh, they continue building sleeping platforms; the rest of the year, Magpie Geese roost in trees or on bare earth (Skutch 1989a, pp. 13–14).

Among passerines, the structures built specifically for roosting rather than rearing young are known from several tropical families, but simpler forms have been found in a few temperate-zone birds. In the Southwest, Curve-billed Thrashers (*Toxostoma curvirostre*) roost much of the year in cholla cacti, but in winter some build stick platforms, while one bird in a pair may roost on the nest previously used for breeding, and the winter platform may later become the base of the next season's nest. In autumn, the thrashers often systematically destroy the unoccupied, much more elaborate roosting nests of Cactus Wrens (*Campylorhynchus brunneicapillus*) within their territory, using some of the material for their own, but they never disturb active wren breeding nests (Anderson and Anderson 1973, p. 170).

Alexander Skutch (1961) popularized the term *dormitory* for a structure birds use for sleeping that is not a nest in which young are raised. He hypothesized that the dormitory-building habit originated in birds that continued using the breeding nest after the young had fledged. He envisioned the initial evolutionary sequence as

- The breeding nest initially being used later by the female that built it,
- Then, in some species used by both parents and sometimes also, in species that have nonbreeding nest helpers, with these helpers, and
- Later, by fledged young, alone or with one or more parent and helper.

From there, Skutch saw the possible extension of the habit to the building of new structures, i.e., dormitories, that were not before and will not later be used as nests. The sequence in occupancy complexity might be

- Built and used by a single bird,
- Used by pairs,
- By parents and still-dependent young,
- By parents and self-supporting young,
- By larger groups.

All these situations are found among birds, although no single family exemplifies all stages. Dormitories likely originated in the tropics, where many passerines build covered nests to protect themselves from rain. Some small birds take weeks or months to construct their nest. Few birds in temperate regions build such nests, because these take longer to construct, and most temperate-zone birds have a short breeding season. In addition, these dormitories would be conspicuous if placed in trees or shrubs that then become leafless. The wrens of higher latitudes that do make dormitories brought this habit with them as their ancestors expanded from the center of wren diversity, Central America and northwestern South America. The globular dormitories of Cactus Wrens are placed in thorny shrubs and cacti where, even if highly visible, they are difficult to approach. In the tropics, some dormitory builders lessen the risk of predation by building several dorms so they can move to another if they see any potential predator when they are approaching one dormitory. Shifting among dormitories also reduces the bird's exposure to parasites that may accumulate in each.

Building a dormitory for sole occupancy is analogous to the habit of most woodpeckers, which excavate a roosting cavity for their own use. The Bananaquit (*Coereba flaveola*) exemplifies the species in which every individual builds its own dormitory; it makes a globular structure with a round entrance facing downward, similar to the breeding nest. Throughout the year, adults sleep singly in these, replacing them as necessary and repelling any other bird, even a prior mate, seeking entry. In a Puerto Rican rainforest, Bananaquit dormitories were shown to save the bird at least one-third of the energy it would have to spend roosting in the open; this was before factoring in the additional thermal benefit the dorm provides by protecting its occupant from the annual average of 3000 mm of rain (Merola-Zwartjes 1998).

The dormitory of the Cabanis's Wren (*Cantorchilus modestus*), also used singly, is a thin horizontal pocket, too shallow to contain eggs, with an entrance on the side (Skutch 1961). More elaborate are the globular dormitories of the Olivaceous Flat-bill (*Rhynchocyclus olivaceus*) suspended from a drooping twig and swinging over clear space as high as six meters off the ground, with an entrance low on the side, sometimes with a spout. They are made of fibrous roots and other fibers and thatched with broad leaves that prevent rain from penetrating (Skutch 1960, p. 514).

Some solo-roosting birds, just like many cavity users, take over nests made by other species. In the llanos of Venezuela, the Plain Thornbird (*Phacellodomus rufifrons*) builds hanging nests, sometimes over two meters in length, primarily of thorny sticks; the nests often have several entrances leading to chambers that are not connected within. At least eleven other species use these nests as nests for themselves or as dormitories, sometimes even when thornbirds are also in residence raising young, with no impact on the thornbirds' success. In this part of their range, Troupials (*Icterus icterus*) nest only in these nests, and roost in them throughout the year; at dusk, they chase one another away from their chosen thornbird nest (Lindell 1996). In southern Africa, some Cape Sparrows (*Passer melanurus*) are sedentary, while others form wandering flocks outside of the breeding season; the latter roost in old nests where they find them, while the sedentary Cape Sparrows build their own individual dormitories (Summers-Smith 1988, p. 70).

Cactus Wrens, conspicuous and confident in the Arizona desert, have been well studied as they build their unmistakable domed dormitories. They always roost singly in one of several dormitories each bird has built, constructing these sleeping nests throughout the year. Their normal materials are grass and other fibrous plant parts, but they incorporate paper, string, rags, chicken feathers, and other debris if they find it. They may begin building at any time of day, but once started, most of the work takes place early in the morning, sometimes before dawn; they may resume, more briefly, in late afternoon. The dorm may be occupied during the first day of construction, but it usually requires seven to ten days to complete. While they never take material from an old dormitory of their own to build a new one, Cactus Wrens do sometimes pilfer the nests of House Finches (*Haemorhous mexicanus*), just as Curve-billed Thrashers more habitually do from the wrens. There is no consistent orientation of the entrance as found in the cavities and burrows of so many other birds. The decisive factor, whether the wren is carrying more material or is unencumbered, seems to be the ease of approach, avoiding the spines of the cactus in which the dormitory is placed (Anderson and Anderson 1973, pp. 24–30).

In his 1961 paper, Skutch gave examples of pairs roosting together in a dormitory. He cited Blue-throated Motmots (in the burrow they used several months later for nesting) rather than any species that has a separate dormitory never used for nesting. When discussing this subject again in 1989, he concluded that what ultimately was used by the motmots as a nest was excavated so far in advance because that was when the soil in banks was most easily worked. The same phenomenon is found in the motmots' Old World burrowing counterparts, the bee-eaters. Whether, in fact, any birds do use non-nest dormitories exclusively as pairs is not known. It is notable, however, that all the birds that sleep together at their nest as mated pairs have nests that are either highly sheltered in cavities, burrows, on cliffs, etc., or are built with a concealing roof and/or entrance.

For Skutch's next proposed stage—parent(s) roosting with newly fledged young—he cites, among others, the Riverside Wren (*Cantorchilus semibadius*) and the White-breasted Wood-Wren (*Henicorhina leucosticta*), which he sometimes found in dormitories with one or both parents and some young. The dormitories of the wood-wren are thin-walled and flat-bottomed, built one to two meters over the ground in much more exposed situations than their substantial and highly concealed nests; both have side entrances. The dormitory is often placed in the fork of a slender sapling, which a predator cannot climb without alerting the occupants (Brewer 2001, p. 189). Among the cavity nesters that also roost as families without building a special dormitory are some woodpeckers, toucans, barbets, swallows, nuthatches, and other wrens (Skutch 1989a, p. 113).

Campylorhynchus wrens, closely related to the Cactus Wren, represent the following stage, in which young roost with one or both parents long after they are otherwise self-sufficient. The Band-backed Wren (*C. zonatus*) builds a bulky globular dormitory that may house parents, helpers from an already fledged generation, and the most recently fledged brood; as many as eleven birds have been found in a single dormitory. The group moves every so often to a new dormitory (Skutch 1989a, p.140).

To accommodate the number of birds in Skutch's final stage, with several families occupying a dormitory, only excavated or natural cavities, or nests deliberately built on a larger scale, could hold such an aggregation. In North America, cavity use is best represented by the Pygmy Nuthatch and White-breasted Nuthatch, which each roost communally for warmth. In the cold and damp cloud forest of Costa Rica, Skutch (1989a, p. 163) observed sixteen Prong-billed Barbets, too many to be of a single family, roosting in a tree hole quite different in shape from the nest cavity a pair excavates for itself. In Argentina, several Lark-like Brushrunners (*Coryphistera alaudina*), a species of ovenbird, work together to build a globular dormitory of sticks in a tree or bush; the entrance is at the top, and as many as eight birds sleep in it (Skutch 1996, p. 150). In East Africa, Black-and-white Mannikins (*Lonchura bicolor*) sleep in their old domed nests, but after these disintegrate, they take over old weaver nests or build dormitories, with several birds participating in the construction of a dorm that can hold more birds than a nest. As many as eleven mannikins, all adults, have been found in a dormitory of their own construction (Van Someren, 1956, p. 477).

Aerial Roosting

Two distinctive features of avian sleep—unihemispheric sleep, in which only half the brain and one eye are at rest, and short bouts between wakenings—likely evolved as antipredator mechanisms, but these have enabled the evolution of an attribute unique to birds, the ability, real or potential, to sleep while in flight. Of the world's more than ten thousand birds, only a few are believed—and still fewer thus far proven—to do this. They are found in very different orders, so the ability is likely to have evolved separately. Of the many birds for which sleep in flight may be a valuable adaptation, especially long-distance migrants, proof will require attaching EEG equipment to their heads and then recovering these devices to read any measurements of slow-wave sleep during the time the birds were aloft.

A flying bird when asleep is likely to be in slow-wave sleep, not rapid eye movement sleep. SWS can be unihemispheric or bihemispheric, while REM sleep is always bihemispheric. REM sleep also substantially reduces muscle tone, which could interfere with flapping flight and possibly also even with gliding or soaring flight. Even if birds in REM sleep could sustain muscles holding the wings outspread, they may need to be aware of the surrounding airspace. Any REM sleep in flight is likely to be extremely brief. In SWS, birds may sleep in flight with half their brain awake when they need to navigate, and they might let the entire brain sleep when constant monitoring is not necessary. The left hemisphere of the brain stores routine knowledge, so this, together with the right eye, is the more likely side to remain awake in flight. Or, birds flying close to others may keep open the eye nearest another bird, as has been found in dolphins, which can swim while sleeping unihemispherically. Finally, just as sedentary birds frequently waken and open both eyes, flying birds may do the same (Rattenborg 2006b).

Swifts

Aerial roosting—without absolute confirmation that the birds are at any time actually asleep—is first and best known from the Eurasian Common Swift (*Apus apus*). Gilbert White (1951, p. 186), in his English village of Selborne, commented in 1774 that "just before they retire whole groups of them assemble high in the air, and squeak and shoot about with wonderful rapidity." He went on to note that the incubating females return to their nest when it is almost dark, but he offered no account of what the others in the flock did. In the mid-twentieth century, David Lack, studying the swifts at Oxford, noted that some never descended or returned to their nests after their upward flight at dusk. As further evidence that they remained aloft, he cited an account from a French World War I pilot flying at night over the Vosges front who encountered widely scattered swifts at 3000 m (Lack 1956, p. 130). More recently, radar tracking has confirmed that swifts are on the wing the entire night.

At the end of the summer day, non-nesting Common Swifts in "screaming parties" rise, often on thermals, from wherever they have been feeding, forming circles that ascend beyond human vision. Radar studies in southern Sweden showed that the column of swifts rises to its highest level, around 3000 m, at dusk, whereupon the birds scatter and then gradually descend to lower elevations, in a mix of flapping and gliding flight, until an hour before dawn. Then, when light levels have increased so that the birds can easily see others, they ascend again. Later, during a rapid morning descent, rather than at peak altitude, the birds form flocks (Nilsson et al. 2019).

The pre-breeding swifts (less than three to five years old) in a colony are the ones that during most of the summer spend the night in the air. Later, they are joined by the newly fledged young birds, which do not return to their nest after they make their first flight, and finally by the breeders, which have roosted in or near their nest. Except during extreme weather conditions—and the debilitation that these may precipitate—no Common Swifts normally land during migration or through their winter in Africa. Radio tracking of birds at their breeding colonies has shown that during the night, swifts aloft orient into the wind, which reduces displacement from their colony, even if sometimes the birds are moved backward in strong winds. With radar, individual swifts were found, however, to drift away from direct orientation into the wind for periods of one to sixteen minutes; these may reflect periods of lowered alertness or actual sleep (Rattenborg 2006b).

Since the swifts are at all times far higher than they need be simply to disperse over airspace or to remain above the level of any aerial noc-turnal predators, the function of these ascents was the focus of a study in the Netherlands using radar to correlate flights with weather condi-tions. It found that the height of the twilight ascent varied evening by evening, being highest on warm nights and fair weather, while the dawn ascent was about 200 m lower. At all times, it was above the level that has significant numbers of aerial insects; only around midnight, when swifts had descended to about 600 m, were they in a zone with many insects. When the swifts ascend at dawn, these insects are descending, so it seems unlikely that foraging is a significant part of the flights. At morning twilight, when the swifts are at 2000 m, they are able to in-crease their visual horizon to 160 km. There, they are able to assess si-multaneously several features detectable only at that time: landscape, light polarization patterns, and stars and magnetic cues useful for orientation, perhaps deployed if wind drift has taken them far from their colony. These may be especially valuable in overcast or foggy conditions, because light polarization is then most perceptible near the horizon. By July, when flocks already include migrants, long-distance navigation as well as relocation of the colony by local birds may be a factor. At 2000 m, swifts can also assess atmospheric conditions and see more distant cloud formations associated with weather fronts, both of which indicate optimal foraging areas for the new day (Dokter et al. 2013).

Recent research has confirmed that the swifts also spend the night on the wing during migration and winter. Micro data loggers used on Swedish Common Swifts found them wintering in West and Central Africa, where they are airborne more than 99 percent of the time. The twilight ascents first observed in Europe occur in Africa, as well as on migration (Hedenström et al. 2016). The same team used similar equipment on Italian Pallid Swifts (*Apus pallidus*) that winter in West Africa and found that these birds also spent more than 99 percent of their time aloft. The few times they landed at night were during periods of the new moon (Hedenström et al. 2019). Alpine Swifts (*Tachymarptis melba*) wintering in Africa tracked with multisensor tags ascended at dusk and, twice as frequently, at dawn, most often during calm and dry weather conditions when no weather front was moving through (Meier et al. 2018). An observational record from Australia of wintering Pacific Swifts (*A. pacificus*) suggests similar behavior; its nesting-season roosting habits have not been documented. On March 10, 1969, a group of about 120 was seen near Frankston, Victoria, far south of their normal range; soon after sunset, the birds formed a dense flock and rose perhaps 1000 m, disappearing from view nine minutes later. The next morning, they were present in the area again, suggesting these were the same birds (Carter 1969).

In the Western Hemisphere, North American Black Swifts (*Cypseloides niger*) wintering in western Amazonia spend eight months aloft. In Brazil, they rise as high as 4400 m at night, while during the day their usual altitude is 470–935 m. The nocturnal flight elevations vary by the phase of the moon, typically 1400 m during the new moon and 2000 m over the eleven nights around a full moon. Their flight is more active when the moon is bright, suggesting that they are taking advantage of the extra light to feed, rather than simply gliding, as they do most of the nights with less moon. As an indication of the importance of moonlight on their nocturnal flight altitudes, during a total lunar eclipse on the night of January 20–21, 2019, swifts then at 3000–4000 m lost altitude rapidly. When the eclipse terminated, they returned to higher altitudes (Hedenström et al. 2022).

Given that most sedentary species of swift roost near their breeding site and that most migrant swifts roost in sites similar to those in which they breed, it may be that aerial roosting evolved among species wintering in landscapes that do not provide enough suitable roost sites. All of the swifts found to roost aerially nest primarily on rock faces and buildings, features that may be scarce or absent in parts of their winter ranges—or perhaps are occupied to saturation by local nonmigratory swifts. Wintering Common, Alpine, and Pacific Swifts reduce their competition with other swifts by generally foraging higher in the air column than resident species (Chantler 1995, pp. 216, 224, 237); aerial roosting may serve a similar purpose.

Anatomical adaptations have made spending as many as ten months aloft no energetic challenge for these swifts. At night, these three species are able to remain for long periods at 1000–3000 m, where oxygen is low, because their high concentrations of hemoglobin enable them to absorb oxygen as efficiently as montane birds living at 2500 m. Only hummingbirds, with their higher metabolic rates, have higher hemoglobin concentrations. Wing shape is also essential: swifts of all species are among the smallest birds to glide, facilitated by their long narrow wings, which can save 72 percent of the energy expended in continuous flapping flight and which allow swifts their low metabolic rate. Compared to similarly sized pigeons and passerines, all with broader wings, their heart is proportionally smaller and their breast muscles are less powerful (Palomeque et al. 1980).

Swallows

Into at least the eighteenth century, it was widely believed that swallows hibernate in winter. Aristotle and other ancients thought they withdrew to caves, cliffs, and hollow trees. Later, mud beneath lakes was seen as the choice of most swallows. This may be because many species of swallow gather in large flocks before migrating and circle conspicuously at dusk over their preferred roosts—wetland reedbeds—finally plunging en masse into the reeds. The occasional discovery of torpid swallows would have constituted further proof. The possibility that these highly aerial birds may roost in the air, as was suspected of swifts long before it was proven, has received only occasional circumstantial evidence.

Early in the morning of July 31, 1975, in Westphalia, Germany, a flock of about two hundred Northern House Martins (*Delichon urbicum*) was seen descending from 1000 m to its colony, where they had not roosted the previous night (Rheinwald 1975). Years earlier, on a spring morning, Julian Huxley saw a flock of House Martins high in the sky at dawn over the Upper Thames valley when there were none at their nests (Buxton 1975). In sub-Saharan Africa, where the bulk of the entire Eurasian population winters, House Martins, unlike wintering Barn Swallows (*Hirundo rustica*), are not known from any roosts, and observational and specimen records from most of their African range are few. They are seen foraging at the same height as Common Swifts, well above the airspace of local swallows (Moreau 1972, p. 120). This is slender evidence for a bird so familiar and abundant throughout Europe; perhaps more conclusive information can come from using some of the same equipment that has revealed so much about swifts.

Frigatebirds

Frigatebirds, much larger than swifts or swallows, have now been fitted with altimeters, satellite transmitters, and data loggers recording brain activity that track the time they are aloft and the time they actually sleep while in flight. Equipment can easily be retrieved when the birds return to their nest—an advantage of working with frigatebirds, in addition to their size. Frigatebirds feed over the tropical and semitropical oceans by dipping to the water surface to pluck fish and by pursuing smaller birds already carrying fish until these drop them, to be caught from the air by the frigatebird. Their feathers are not waterproof, and their tiny, webless feet are not suited for swimming, so they must remain in the air. Their long wings—proportionally wider than those of albatrosses—and small body mass enable them to ride on thermals, even the relatively weak ones that the trade winds at low latitudes maintain over the sea both day and night. Within this atmospheric belt, frigatebirds are able to spend weeks or months in the air. As with the swifts, frigatebirds have evolved additional features that enable them to live on the wing. Their pectoral muscles have a layer not found in other birds that assists both in soaring and sailing flight and enables greater maneuverability (Kuroda 1961).

A study of Magnificent Frigatebirds (*Fregata magnificens*) nesting on an island off the coast of French Guiana fitted some birds with continuously recording altimeters and others with satellite transmitters. It found that the birds rise as high as 2500 m, where the air temperature is 10–12°C, compared with 27–30°C at sea level. Their flight is a succession of rises and descents, gliding down in the direction they are traveling and coming near the sea surface only by day. At night, their ground speed is 8.6km / hour; during the day 10km / hour (Weimerskirch et al. 2003). Subsequent work with Great Frigatebirds (*F. minor*) nesting on Europa Island, between Africa and Madagascar, equipped birds with electrocardiography and GPS devices and found that some birds rode the trade winds at the edge of the Indian Ocean doldrums for more than two months, traveling as far east as the coast of Sumatra. They flew at altitudes from the sea surface to 3000 m, occasionally to 4120 m, where the air density and oxygen available are almost half what they are at sea level and the temperature is 0°C. Most flight, however, was between 0–600 m in elevation. The ascents to higher levels peak during the first hours of night, done without flapping, and mostly within cumulus clouds that provide updrafts lifting the birds at five meters per second. From 4000 m, a frigatebird can glide more than 60 km, compared with the 17 km if it started at the base of a cumulus cloud. It is during the soaring, and perhaps also the gliding periods when the wings are motionless, that frigatebirds are likely to sleep. These motionless periods are relatively short, averaging two minutes and never longer than twelve minutes, which are within the range of single sleep episodes of birds on the ground (Weimerskirch et al. 2016).

Definitive evidence of sleeping in flight came from Great Frigatebirds on Genovesa Island in the Galapagos. The birds carried lightweight sensors recording EEG, data loggers measuring triaxial acceleration, and GPS loggers, none of which impeded their flight or other behavior. These birds, all females caring for small chicks, had foraging trips (while their mates replaced them on the nest) averaging 5.8 days, with one extending to ten days and covering 3000 km. During the day, the EEG sensors showed waking activity in both brain hemispheres. Shortly after sunset, however, they showed slow-wave sleep when the birds were soaring or gliding. This was usually unihemispheric, but it was sometimes bihemispheric, with no impairment of flying ability. Compared with the amount of bihemispheric sleep frigatebirds have while on land, there was less in flight (Rattenborg 2017).

Most sleep was when the birds were soaring in a circling pattern, and never while flapping. As they turned in a circle, the frigatebirds slept with the eye in the turning direction open and the opposite brain hemisphere awake; this may reduce the chance of any collisions, although frigatebirds are thinly dispersed over the ocean and it is unlikely any other is in the same thermal. The birds had no preference for circling rightward or leftward. There were also brief episodes of REM sleep, averaging 4.9 seconds, compared with 5.9 seconds when on land, and only 3.5 percent of total sleep time versus 8.2 percent on land; during REM in flight, the head dropped, as it does on land. Each total sleep episode was also briefer—averaging 12 seconds (compared to 52 seconds on land), with the longest 348 versus 1,212 seconds—and the total sleep time per day was only 0.7 hour, while on land it was 12.8 hours, when the birds were at their nest. The birds are evidently able to function with far less sleep than their time aloft makes available; when they returned to their nests, their initial sleep was more intense than later. Whether frigatebirds sleep more on the much longer flights they make when not tending a nest remains to be tested (Rattenborg et al. 2016).

Other Birds

Now that definite sleep has been shown in flying Great Frigatebirds, the necessary technology may be adapted to other birds that spend long periods on the wing. The challenge, however, is in recovering the equipment that has collected data on the flying bird—easy with a frigate-bird returning to its nest, especially in the Galapagos, where frigatebirds are likely tamer than in many other colonies. This is more problematic with long-distance migrants and with seabirds that spend years away from their natal colony. The bird believed to spend longest at sea without landing is the Sooty Tern (*Onychoprion fuscatus*), which, like frigatebirds, rarely touches the water and travels far on foraging trips. Young Sooty Terns may not return to nest until they are ten years old, spending all those years over the tropical ocean. The pre-breeding birds from the colony in the Dry Tortugas, beyond the Florida Keys, are believed to spend at least two to five years off the coast of West Africa, where upwellings make fishing easier for inexperienced birds than in the seas nearer Florida where adults winter. When flotsam, sea turtles, and other floating objects are not available, the birds likely remain aloft (Schreiber et al. 2002). More definitive evidence comes from adult Sooty Terns from a colony on Bird Island, Seychelles, in the western Indian Ocean. Birds were tracked with Global Location Sensors while at sea during their nearly seven-month nonbreeding season. They were found to spend less than 4 percent of their time in contact with seawater and never to land there at night (Jaeger et al. 2017).

On a much shorter timescale, birds that migrate long distances for several days over the sea and cannot land might also get some form of sleep while flying. These include several plovers and sandpipers that have nonstop flights over the Pacific and Atlantic oceans, with that of the Bar-tailed Godwit (*Limosa lapponica*) lasting as long as eight days from Alaska to New Zealand, as well as the Amur Falcon (*Falco amurensis*), which flies for more than five days over the Indian Ocean between south-west India and Somalia, and the Blackpoll Warbler (*Setophaga striata*) in autumn traveling for three days over the Atlantic from northeastern North America to Caribbean islands on its way to South America. Many other small birds have migratory flights of one to three days crossing the Gulf of Mexico and the Mediterranean Sea and Sahara Desert. Results with Great Frigatebirds show that they are awake when flapping, and the flight of all these migrants requires almost continuous flapping, suggesting a lower probability that they are able to shut down any part of their brain (Rattenborg 2017).

Whether or not these migrants completing long overwater flights have slept en route, observations of some birds that have just landed show that sleep is their first priority. Northbound Red Knots (*Calidris canutus*), Bar-tailed Godwits, and Black-bellied Plovers (*Pluvialis squatarola*) arriving on the island of Texel, The Netherlands, likely in the air since leaving West Africa, go to sleep in minutes on the mudflats. A few hours later, they fly several kilometers to wetlands in the Wadden Sea to feed. Pink-footed Geese (*Anser brachyrhynchus*) landing in Scotland from Iceland first drink copiously then sleep standing on one leg in shallow water. They do not begin feeding until the next day (Schwilch et al. 2002).

Water

Birds that routinely swim also regularly roost in water. Some use lakes and bays where staying far from shore puts space between waterfowl and any ground predator and makes aerial predators easy to spot when these are still at a distance. Other birds roost at sea, where highly pelagic birds rest between dives or foraging flights and during the night. Thanks to data loggers that can distinguish their flight from periods of immobility, albatrosses, which were hypothesized to sleep on the wing, are now known to rest on water for much of the night, especially when there is little moonlight. Other birds, however, are occasionally forced to land on water when far from their usual roost sites.

On two January nights, as the heavy fog came over the Dutch Wadden Sea with the advancing tide, a group of ornithologists on a boat headed for one of the islands encountered shorebirds of several species floating on the water or taking off and fluttering when a spotlight startled them. Red Knots and Dunlins (*Calidris alpina*) were on the water, and Bar-tailed Godwits, Curlews (*Numenius arquata*), and Oystercatchers (*Haematopus ostralegus*) were heard around the boat. On one of the nights, the boat was surrounded by these birds about one kilometer from their nearest island roost; three nights later, the birds were on the water two to three kilometers from shore. It is likely that the fog overtook the birds as they were returning from their feeding flats when these were submerged by the tide. The birds became disoriented and did not continue, no longer sure of the correct direction, when visibility was 100 m and the winds were strong (Piersma et al. 2002).

In a completely different situation, a flock of three thousand Bar-tailed Godwits and a few Red Knots were found in a dense flock on the water off the coast of Guinea-Bissau on a February afternoon during the hours of high tide. They spent at least ninety minutes on the water. The air temperature there was 38°C and the water temperature 32°C. On the sand flats where these birds sometimes roost, the temperature at soil level was 60°C. Similarly, Spotted Redshanks (*Tringa erythropus*) have been seen roosting at sea for several hours off the coast of Ghana during hot weather; this species is also known to roost on water in the Wadden Sea (Piersma et al. 2002). These observations suggest that some shorebirds may, on long overwater flights, land and rest without becoming waterlogged.

Summary

Some birds go to great lengths to secure roost sites that provide substantial protection from predators or the elements—much more protection than most birds receive when sheltering in vegetation. These birds seek out or make for themselves a sleeping place that provides additional warmth, concealment, or structural defense. Most are small birds that roost alone or in small groups, lacking the benefits of flockmates alerting for danger. Extreme cold or scarcity of safe sites may also drive larger birds to unusual sites. Cavities in trees, burrows in the ground, and tunnels in snow all meet these needs, each forming a protective layer hollowed out within some other material. Other birds build elaborate structures with external defenses; these are used by the builder, a pair, or an extended family group as a communal roost. Still other birds go to the opposite extreme from any fixed site; they take to the sky, spending the night aloft beyond where anything would pursue them.

Cavities in trees provide concealment, protection from wind and rain, and natural insulation that can be augmented by a bird's own body heat. Not all cavities, however, are equally useful. Woodpeckers make their own, often several, to be used over the nonbreeding season. These are custom built to the bird's own size, with openings oriented to sunlight and dry winds, and in wood, dead or living, with the air circulation appropriate to the season. Some other birds, including woodpecker relatives the barbets and toucans, also make cavities, albeit in softer, sometimes rotting, wood. Many smaller birds incapable of making their own cavity seek out unused woodpecker holes and natural ones to use as both roosts and nests.

Where trees are scarce or absent, some birds dig holes in the ground or use those made by other birds for nesting as well as ones made by other animals. As with tree cavities, these may be significantly warmer than the surrounding environment. Most burrows are configured to prevent flooding, but this, as well as collapse, may sometimes happen. Because burrows require more energy to excavate than tree cavities, birds that use them usually have fewer and use each for more of the year, so the burrows hold more parasites than tree cavities, which most birds keep clean.

More ephemeral are tunnels in snow, which some birds make where snow is regular, tree holes scarce, and ground frozen. The most significant users of snow are various grouse, which in any case could not excavate a cavity in wood or soil. They usually create the tunnels by flying into a snowbank. At high northern latitudes, where most of the winter has little or no sunlight, grouse may spend nearly all the twenty-four hours in their snow burrow, where they generate a microclimate much warmer than anything available in an exposed situation. Smaller birds regularly or in extreme situations resort to snow burrows, usually for much shorter durations.

In both tropical lowlands and high in the mountains, caves are used as roosts by some birds that nest elsewhere. Caves have a very stable temperature regime, making them attractive in tropical forests as a place to shelter from the heat of midday and at great elevations to take refuge from the extreme cold of night. In the Andes, birds as varied as hummingbirds, owls, and geese roost in caves at night. Still higher, where there are no caves, crevasses in glaciers also serve as roosts. In other mountainous areas, abandoned buildings and mineshafts serve the same purpose.

Some birds sleep on their nest as they are constructing them—when occupancy may provide thermal benefits or prevent rivals from using the site—as well as through the nesting cycle and beyond. Other birds build special structures for roosting that are never used as nests. In wetlands, where aquatic birds have few places to spend long periods out of water, some coots and geese pull vegetation together to create mounds over the waterline where they can rest or sleep, with or without their young after these have left the nest. Because it is time-consuming to build special roosting structures in addition to nests, this habit is found in nonmigratory species, among passerines and other smaller birds, mainly in the tropics. It likely evolved from nest building, with the actual nest then used by the builder(s) and perhaps their fledged young. Since these nests may not be very durable, some birds developed the habit of building replacements for use in months before the subsequent breeding season. Today, these "dormitories" are used, in different species and families, by single birds, by pairs, by parents and still-dependent young, by parents and self-supporting young, and by larger groups of unrelated birds. Other birds simply move into the unused—or even the occupied—nests and old dormitories of other species.

Some highly aerial birds take an entirely different approach to roosting—they spend the night hours in flight. Some swifts, swallows, nonwaterproof seabirds like frigatebirds and certain terns, are known to do this, and many highly migratory birds that have flights of a few days without landing may also sleep in some form while aloft. Thus far, however, the data loggers recording brain activity that can definitively identify sleep during flight have only been applied to Great Frigatebirds; these showed that the birds did have short bouts of SWS, both unihemispheric and bihemispheric, and even small bits of REM sleep.

Finally, birds that routinely swim may roost on the water rather than land. Ducks and other nearshore feeders move farther from land to sleep, beyond the distance any predator would swim or fly to reach them. True seabirds spend much of the night on water, especially when lack of moonlight prevents them from foraging or flying. Even shorebirds sometimes rest for long periods on water if fog prevents them from finding land or, during the day, if the sandflats where they normally wait out high tides are intensely hot.

5

Social Aspects of Roosting

Even for birds that do not roost in flocks at any time during their annual cycle, roosting behavior has social aspects, when birds pair up, raise young, and when those young in turn are first seeking their own mates. Only the most solitary or pugnacious birds, such as hummingbirds, whose sole nonaggressive social interaction is a brief courtship and insemination, do not associate together at some level when roosting. For the many birds that maintain their pair bond through the breeding season, through the year, or over a lifetime, sleeping in some degree of proximity is one of the behaviors that maintains the relationship, and it may ensure that the relationship is not disrupted by a rival. Some birds continue to roost as families even after the young are otherwise self-sufficient. Among some birds that nest solitarily and territorially, the birds not actually on the nest, incubating eggs or brooding young, may join others in a territorial no-man's-land to roost. When the nesting season is over, many additional once-territorial species then change their social behavior to roost in larger aggregations, even if they do not spend the day feeding in groups.

Pairs and Pairing

Alexander Skutch, in his many nighttime searches for roosting birds in Panama and Costa Rica, rarely found any sleeping alone—while acknowledging that finding solitary birds in thick foliage in a dark forest using only a flashlight would yield disproportionately low results. He had more success scanning trees in gardens and at clearing edges, where over the years his light picked out flycatchers of several species, each with a mate perched within one or a few meters. He found pairs of tanagers sleeping even more closely to one another, only centimeters apart, and some trees held several pairs of different tanager species, all in pairs (Skutch 1989a, pp. 31–34).

Maintaining the Pair Bond

Males of many territorial species escort their mates to a roost site, which they may or may not use themselves. At the edge of a much-traveled road in São Paulo, Brazil, a pair of Sayaca Tanagers (*Thraupis sayaca*) roosted for at least several nights 4 m up in a 6-m bush, with the female going to a branch in a clump of leaves, followed by the male, who first perched about 20 cm below her and then moved to the same branch, disturbed neither by the observers nor by heavy trucks. A nest that might have been theirs from the previous breeding season was on another branch nearby, so this pair's fidelity to the site, despite the disturbance and when others were available, could have been to prevent another pair from claiming the bush in anticipation of nesting there (Oniki and Willis 2002). The same Brazilian observers watched a pair of Burnished-buff Tanagers (*Tangara cayana*) come for several evenings to a sapling on the campus of the Universidade Estadual Paulista, also in São Paulo State, where the female entered a dense cluster of leaves 10 m above the ground to roost while the male moved about and then departed. One morning, the male arrived at 6 a.m. and the female emerged at 6:12; the pair then went off to forage together (Willis and Oniki 2003).

Among cavity roosters, escorting by the male, who then awaits his mate in the morning, is widespread. During the breeding season in Scandinavia, Siberian Tits (*Poecile cinctus*) only roost at 11 p.m. and rise again fifty to seventy minutes after sunrise, at 2:30 or 3 a.m. Males fly with their mate to the nest cavity, where she roosts; the male then waits a few minutes nearby and goes on to his own roost site on a branch in a tree, near its trunk. At dawn, he may rise thirty minutes before the female emerges, waiting close by for her. At the very beginning of the nesting season, before the female has begun laying and through the first few eggs, the male roosts even closer to the cavity, often in the same tree (Hailman and Haftorn 1995). This is likely a method to ensure that she does not receive or respond to the attentions of any other male. Similarly, male Great Tits (*Parus major*) retire later than their mates, and in the morning they perch near the nest cavity and sing, sometimes for as many as forty-five minutes, before the female emerges (Kluijver 1950). This may likewise be a form of mate guarding, while simultaneously announcing to neighboring males that he is by her side to defend her, since, in many birds, including Great Tits, extra-pair copulations occur most frequently in twilight or before sunrise (Schlicht et al. 2014).

During the breeding season, male Eurasian Nuthatches (*Sitta europaea*) routinely escort their mate to her roost; they roost singly, usually in cavities, but in the summer sometimes elsewhere in trees (Matthysen 1998, p. 76). Male Brown-headed Nuthatches (*S. pusilla*) begin roosting together in the nest cavity two weeks before laying begins; this seems yet another strategy to prevent extra-pair copulations. Females typically enter the cavity earlier and depart later than males. When the young have fledged, the adult pair continues roosting side by side, but on branches, while the not-yet-independent young roost together on another branch (Norris 1958, p. 270).

A pair of the much larger cavity rooster, the Pileated Woodpecker (*Dryocopus pileatus*), was found on cloudy autumn days in Maine to arrive together at the female's roost; after she entered, she sometimes watched the male from the entrance while he foraged nearby. On two evenings, he followed her to the entrance and preened there for several minutes before flying to his own cavity. Another pair behaved the same way, but with the female following the male to his roost. On one morning, the male in the first pair flew to the tree with the female's cavity, vocalized, and tapped on the tree for fourteen minutes as she watched from the cavity entrance. Another morning, this male drummed near his own roost tree and the female responded by drumming from her cavity, then flying to his, where he joined her, entering the cavity and tapping from the inside while she tapped from outside. In these activities maintaining the pair bond between nesting seasons, the birds' roost sites seem to play a role, even though Pileated Woodpeckers always excavate a new cavity for nesting the following spring (Kellam 2003).

Birds that nest or roost on and near the ground are not as limited as cavity users, and their roosting behavior, as well as its social aspects, may change more often during their annual cycle. The Sichuan Hill Partridge (*Arborophila rufipectus*) of southwest China lives in subtropical montane forests, where it feeds on the ground. During early spring, males establish territories and roost solitarily a few meters off the ground. When they have formed pairs, males and females forage together and roost within 2.5–3.5 m of each other, but on different shrubs. Here, too, the males' close association with the female just before egg-laying is likely to guard her. Slightly later, when the female begins incubation on a nest on the ground, the male no longer roosts nearby, perhaps thereby reducing the potential notice of nocturnal predators. After hatching, however, the male stays near the brooding female and roosts on the ground close by until the chicks are nine to thirteen days old. Then he separates from the brood both by day and by night, returning to roosting alone in shrubs until the following spring. Although not as limited as cavity users, at all seasons the hill-partridge is selective about roost sites. Of the thirty-two tree and fifty-two shrub species found in its habitat, it roosts in only seven species of shrubs, all of which have particularly dense foliage (Liao et al. 2008).

Among birds with pair bonds that extend beyond a single breeding season, roosting together throughout the year is one of the ways the bond is maintained. Pairs of long-lived, highly sedentary African Fish Eagles (*Haliaeetus vocifer*) disperse each day to feed in different localities—the better to defend their expansive territory—but return nightly to roost on the same branch of a big tree away from the water's edge, using the same site for long stretches (Brown 1980, p. 40). Rooks (*Corvus frugilegus*) and Jackdaws (*C. monedula*) both form lifelong pair bonds; they roost communally, sometimes in mixed flocks, and disperse every day to feed. Whether or not pairs stick together through the day, by evening more of the birds in the flocks returning to the roost are in twos, flying more closely together than to any of the other birds. It has not been proven, but it seems likely that these are pairs (Jolles et al. 2013). On arrival at their joint roost, Rooks generally settle in taller trees, Jackdaws in shorter ones; there is constant movement and noise as the birds select perches and, perhaps, are vocalizing to find their mates in what is now often darkness. By the time they are all settled and quiet, nearly all birds are in pairs (Coombs 1976, p. 104). Pairs of Common Mynas (*Acridotheres tristis*) are known to spend the day together, whether feeding by themselves or in larger aggregations; at night, when their roosts may constitute a few tens of birds or several thousands in a group of trees, pairs rest close to one another (Feare and Craig 1999, p. 159).

Forming a Pair Bond

Communal roosts are also an efficient way for young birds to find a mate for the coming breeding season. In South Africa, the territorial African Black Duck (*Anas sparsa*) lives on rivers, where it roosts on the shore. At least since the building of irrigation dams on these rivers, many have taken to gathering at the end of the day on the larger bodies of water created by the dams, where less vegetation at the edges has made for safer roosting. Between sunset and darkness, they interact in groups, usually of two to four, sometimes with overt fighting and aerial chases. By night-fall, the birds settle quietly on the beaches created by the dams. At these gatherings and through these interactions, young Black Ducks are thought to find mates (Siegfried et al. 1977). In the Everglades, a roost of Bald Eagles (*Haliaeetus leucocephalus*) in pine trees far from foraging areas and providing no climatic benefits is populated during the breeding season mainly by subadults; here, some have been seen forming pairs with other young birds and with single adults (Curnutt 1992).

For adults also, communal roosts are an important site for pair forma-tion, both during the breeding season and earlier. On late winter morn-ings, male Wild Turkeys (*Meleagris gallopavo*) display in roost trees before the flock descends to the ground to feed; more occasionally, they copu-late in roost trees at dusk (Buford et al. 1994). Royal Tern (*Thalasseus maximus*) courtship occurs at winter and spring roosts far from breeding grounds, as well as at the colony's pre-breeding roost, which is usually within 100 m of the future nesting area (Buckley and Buckley 2002).

Among cavity-nesting birds, the cavity and the benefits and privi-leges of roosting in it may be part of courtship and pair formation. Male Purple Martins (*Progne subis*) returning in spring to colonial martin houses establish a territory of one or more compartments. Females se-lecting a mate seek to sleep in the same compartment; some males are more willing than others, so at the early stages of pair formation, some females sleep in separate compartments within the male's territory. Sharing a compartment may begin as many as six days before the first egg is laid. Copulation likely occurs there, since it is rarely seen during the day. Some pairs sleep together during egg laying and incubation, reducing the likelihood of extra-pair copulations, and continue in the nest with the nestlings until these are at least fourteen days old. But males that sleep in a separate compartment during incubation continue to do so after the young hatch, and some pairs reunite in a single com-partment later. After the young fledge, nearly all martins roost in trees, while still spending part of their day at their houses as premigratory flocking begins (Brown 1980). At some colonies, however, adults con-tinue to roost in their compartments longer, defending them against the newly fledged birds from other colonies that arrive at the end of the season, seeking to occupy the compartments, perhaps in anticipation of returning to nest the following year (Morton and Patterson 1983).

In some species, roosts have a set of hierarchies, with birds of different ages and mating status in different sections. Red-billed Choughs (*Pyrrhocorax pyrrhocorax*) do not breed until they are two or three years old. In Spain, younger birds roost communally throughout the year, with nonbreeding birds socially segregated according to their age and breeding prospects. Breeding pairs roost together year-round at their nest site, so the communal roosts are populated only by unattached birds of various ages. The main roosts are in areas of higher breeding density and include more first-year birds. Subroosts, holding only 20–25 percent as many birds as the main roosts and at various distances from them, contain more choughs of breeding age and are in areas with more available nest sites. When they are older, the young nonbreeding individuals first observed in main roosts move to subroosts, while the converse is rare. At the subroosts, more choughs acquire mates than in the main roost. The males here are in poorer condition than those already breeding, while the females at the subroosts are in better condition than those in the main roost. The choughs here pair most often with roost-mates and also with widowed territorial breeders. And, given that the subroosts are in areas with more available nest sites, more of the birds that mate there acquire territories closer to that roost than do the few birds at the main roost that form pairs. By the beginning of the breeding season, these subroosts are empty. Thus, the different roosts act like a social conveyor belt, with the main roosts receiving new young birds, which later graduate to the subroosts, where they find a mate, depart to establish a territory, and do not return unless one in the pair dies (Blanco and Tella 1999).

Young

Young birds just out of the nest are perhaps at the most vulnerable stage of their lives due to their lack of experience and still-developing flying ability. At night, they are even more vulnerable. But after they leave the nest, most young birds roost in the same sort of place as their parents, and few seem to be taught how to find a place that provides the first two priorities of roosting sites: concealment from predators and protection from the elements. Only a relative few birds are known to give their young any guidance in roost site selection—and these are all sedentary species for which the lessons taught to young birds will apply for the rest of their lives in the same area. The young of migratory birds, except those few species that travel with their parents, must determine for themselves the features of a good roosting place at stopover sites and winter destinations.

Young Roosting with Their Parents after Fledging

In some ground-nesting birds with precocial young that can feed themselves from the time they dry off on emerging from their egg, one or both parents continue to look after them for several weeks. In Pennsylvania, Ruffed Grouse (*Bonasa umbellus*) females brood their chicks for about five weeks, evidently the time it takes the young to develop complete thermoregulation as well as sustained flight. During those weeks, the family sleeps on the ground, with the female spreading her wings over the young to help insulate them. The usual site is at the base of a tree, which may give some protection from wind, as well as physical and scent barriers to mammalian predators. Although the young are capable of short flights at fourteen days, it is not usually until three weeks later that the female leads them to trees, where they then roost in thick vegetation that provides concealment and insulation. They remain with their mother until twelve to fifteen weeks old (Tirpak et al. 2005).

Young geese and cranes typically remain with their parents into their first winter, migrating with them and continuing to stay close until the adults prepare for the spring migration. Among Common Cranes (*Grus grus*) wintering in Spain, some pairs with young of the year maintain a family feeding territory during the day, while other adults, with or without young, spend their days in flocks that roam among feeding areas. In the evening, however, the territorial families join the communal roosts used by all the cranes in the area (Alonso et al. 2004). For them, as for geese that winter in similar open habitats, a few birds by themselves would be both conspicuous and vulnerable, while a large flock, although even more conspicuous, always includes some vigilant birds all through the night.

Parents Guiding Their Young to a Roost

Most reports of young birds guided to roost sites concern cavity nesters and species that build domed nests, which provide much of the same security. Blue-throated Bee-eaters (*Merops viridis*) of Southeast Asia lead their young back to the burrow in which they were born for at least ten nights after fledging, while the parents then roost elsewhere; for them, safety rather than warmth is likely the primary motivation (Fry 1984, p. 113). Young Pygmy Nuthatches (*Sitta pygmaea*) continue roosting in their nesting cavity after they have fledged and their parents remain with them. At one nest in Marin County, California, young twenty-four days old and a few days since fledging were led back to the nest about ninety minutes before the parents joined them for the night; the next morning, the young remained in the nest thirty minutes later than their parents. At another nest where parents went to roost at the same time as their young, the parents waited until all their young had entered the cavity before going in themselves (Norris 1958, p. 267). While tits are all cavity nesters, only one species is known to bring its fledged young back to the nest. A pair of East African White-bellied Tits (*Melaniparus albiventris*) was once observed leading its newly fledged young back to the nest cavity, where both parents also roosted (Van Someren 1956).

After young House Wrens (*Troglodytes aedon*) in Central America leave the nest, the family may return to it to sleep or use another cavity nearby. Parents show their fledglings how to enter a roost. The adults go in and out several times until the young follow them. When the parents have a second brood, young of the first brood may still be sleeping in the nest and may help feed the new brood (Skutch 1940). Several other cavity-nesting wrens have similar habits, while Cactus Wren (*Campylo-rhynchus brunneicapillus*) pairs lead their fledged young back to their nest or to one of the dormitories the adults have previously built for their own use. Adults and young do not sleep together, but the parents continue leading their young to various dormitories within their territory. At fifty-two to seventy days, the young split up, each to a separate dormitory, and at about three months, they each start building their own (Skutch 1989a, p. 118).

Green Woodhoopoes (*Phoeniculus purpureus*) in East Africa very conspicuously lead the young to roost. Woodhoopoes live in groups of up to seventeen birds, with one breeding pair and several helpers from prior generations of young. All flock members continue to care for the young after they fledge. For about two weeks, as many as nine may work as a group to entice or guide the new fledglings to that night's roost cavity in a tree. About an hour before dark, they fly as a group to a tree, giving excited calls until the fledglings join them. Repeating this, they gradually reach the roost tree. There, the young are led to the hole, with an adult entering first. Other adults bring food until the young settle down. The entire process may take thirty minutes, ending well before dark. The adults preen or resume feeding; afterward, they either enter the cavity with the young or going off to their own cavity (Ligon and Ligon 1978).

Perhaps because it is more difficult to observe, there are few accounts of adults leading their young to exposed roost perches. Puerto Rican Todies (*Todus mexicanus*) nest in burrows, but when the young fledge, they never return to them. Parents lead them to a branch in foliage several meters off the ground, where each young bird sleeps alone (Kepler 1977). Skutch (1997, p. 112) found that when some young tropical flycatchers leave the nest, they follow their parents and perch together, bodies touching, waiting to be fed. They sleep the same way, with bodies pressed together in a row and a parent nearby or in the same row. Among African and Asian honeyguides, which all deposit their eggs in the nests of other species, the young of the Eastern Green-backed Honeyguide (*Prodotiscus zambesiae*) of southern Africa stay with their foster parents for a few days after fledging—having earlier killed any nestmates hatched from eggs that the mother honeyguide did not remove when laying her own—and roost on a branch between the two foster parents (Short and Horne 2001, p. 111).

Long-term Family Roosts

At the other end of the spectrum, long-term family roosting is found in birds like Andean Condors (*Vultur gryphus*), which do not attain maturity until about eight years old. Their cliff roost sites tend to hold individuals that are closely related, although their nests are more widely scattered. Young males eventually disperse much farther from their natal home than young females. The benefits to the birds in these family aggregations may include defense by the dominant adult males against unrelated intruders and sharing of food resources among birds that are genetically linked; individuals from the same roost fly in different directions to search for food, but they can watch each other from the air to see if one is descending to a carcass. When a carcass is found, the family members can dismember it more rapidly if they do not have to defend it against unrelated birds. If it provides food for more than a single day, any naive family members can follow the knowledgeable ones when these depart the roost (Padró et al. 2019).

Among Black Vultures (*Coragyps atratus*), for which roosting sites, usually in trees, are more common than for Andean Condors with their specialized cliff requirements, individuals move among several roosts within their foraging range, but they usually settle each night with their mate and offspring from recent years. Vultures known to be related often do mutual preening at the roosts, while only unrelated birds have been seen fighting. This long-term family cohesion, reinforced at roosts, reduces conflicts there and increases the likelihood that family members will search for carcasses over adjacent areas the next day. When a carcass is found, there will be fewer disputes and the family can dominate other vultures seeking access to it. This is particularly beneficial to young vultures with less foraging experience and subordinate status in any conflict (Parker et al. 1995).

Huddling

Huddling together for warmth—known as "social thermoregulation"—likely originated in the nest, where young birds, whether brooded by a parent or left on their own, benefit from the body heat generated by the others around them until they have developed the ability to maintain their own body temperature. On cold nights, and sometimes even on warm days, some birds huddle with their family, with unrelated individuals of their species, and, when few other birds are available, even in mixed groups with other species. Among some cooperative breeders, in which young of previous years remain with their parents for at least one additional breeding cycle, huddling occurs regularly during the day, when it may serve as a form of group cohesion reinforcement.

Recent studies of olfaction in birds have shown that scents derived from uropygial secretions of preen oil and their associated microbes, especially in females, enable birds to recognize another bird's species, sex, breeding condition, individual identity, and kinship. In addition, in species that have been tested, both adults and young can recognize the scent of their own nest (Whittaker and Hagelin 2021). While this has not been studied experimentally in huddling birds, it seems likely that scent helps enable members of the group to recognize the hierarchies that are important in some huddling species as well as to identify any unfamiliar or unrelated individual attempting to join the huddle, whether on a branch or within a cavity or dormitory. The number of cooperative breeding and huddling species that have been observed in mutual preening suggests that this behavior further homogenizes the scent of related birds, making them all the easier to identify and creating strong cohesion within the group.

Sites

As with the wrens that return to their nest or to a dormitory with their young, other birds also use the nest or a similar place to huddle in a group. Cavities, with their insulation and concealment, are widely used by birds that huddle—but only by species that nest in cavities, so they are already comfortable entering them. The bodies of the several birds together elevate the cavity temperature, thereby reducing their own energetic demands. The exposed surface area of each bird is also reduced, further decreasing heat loss. And since the temperature of the plumage of the huddled birds is similar, there is little or no drop in body temperature as would happen if heat flowed from one bird to a cooler one adjacent.

Observations of birds huddling on branches in the lowland tropics suggest that some species are extremely sensitive to the nightly drop in temperature, or that thermal benefits are not always the motivating factor. Bee-eaters in both tropical and temperate zones routinely sleep in dense rows during warm nights as well as cold ones. In Australia, eight Rainbow Bee-eaters (*Merops ornatus*) were found roosting facing in the same direction, with bodies touching, on a warm February night when the temperature was 25°C (Warham 1957).

ENCLOSED SPACES

Cavities are so desirable that birds may cross their neighbors' territories to reach one. British Eurasian Wrens (*Troglodytes troglodytes*) may crowd with thirty or more birds in a cavity; some travel more than 1.76 km to a preferred site (Armstrong 1955, p. 281). Pygmy Nut-hatches, which roost as a family in their nest cavity for some time after the young have fledged, later merge into larger aggregations. During the nonbreeding season they sleep tightly packed in cavities with other families. In Colorado Springs, Colorado, an October flock was esti-mated at 150 birds, with about 100 roosting in a natural cavity in a pine and the balance in smaller cavities in the same tree (Knorr 1957).

In the tropics, some cavity-nesting birds also sleep in larger groups when not nesting. The White-eared Barbet (*Stactolaema leucotis*) of eastern and southern African highlands roosts in groups of up to eleven in a cavity, and the Prong-billed Barbet (*Semnornis frantzii*), of Costa Rican and Panamanian cloud forest, roosts with as many as nineteen in a cavity. During the day in the nonbreeding season, Prong-billed Barbets are in groups of not more than five, so their tolerance at night of less familiar birds in close contact indicates a need for the benefits of warmth generated by a larger assemblage. It is likely that cavities holding five birds are commoner in the forest than those holding nineteen, but these smaller ones are not preferred. In the forested lowlands of Mexico, Ecuador, and Venezuela, the Collared Aracari (*Pteroglossus torquatus*) roosts in old woodpecker holes, with up to six birds tightly packed, their tails over their back pressed against their nape (Short and Horne 2001, pp. 139, 318, 396).

An enclosed, well-insulated dormitory provides some of the same benefits. In South Africa, where winter nighttime temperatures frequently drop to below 5°C, Bronze Mannikins (*Lonchura cucullata*), only 8–13 grams, live in small flocks of up to thirty birds after the breeding season; like several other mannikins, they build a communal dormitory in which they roost, huddled together (Calf et al. 2002). Some birds, like the Black-tailed Gnatcatcher (*Polioptila melanura*) of the Southwest, which in winter occupies old nests of Verdins (*Auriparus flaviceps*) in groups of as many as sixteen, regularly use the better insulated nests made by other species (Walsberg 1990).

ON THE GROUND

Other huddling birds gather together on the ground. Coveys of North-ern Bob-white (*Colinus virginianus*) assemble in a compact circle under some sheltering vegetation, with their tails together and their heads outward, so they can detect danger from any direction. The first two birds to reach the sleeping spot stand pressed together, facing the same way; others join next to them, forming an arc that eventually meets in a circle. Late arrivals wedge themselves between others. If the quail are disturbed during the night, each flies in the direction it has faced, elimi-nating the danger of collision in the dark (Skutch 1989a, p. 23). In sea-bird colonies on islands without ground predators, birds not on a nest may roost in close proximity, sometimes with several touching one an-other, like the Red-footed Boobies (*Sula sula*) of all ages and in all color phases on Raine Island at 11°S off the Great Barrier Reef. Here, however, warmth does not seem the likely reason for physical contact (Warham 1961).

On a much larger scale, Emperor Penguins (*Aptenodytes forsteri*) in Antarctica huddle in groups that may include more than a thousand birds. Birds join huddles from nearby loose aggregations; as others join around them, the situation of each bird shifts from the edge toward the

Emperor Penguins in Antarctica move in and out of massive huddles to warm up and then to cool off. (Photo credit: André Ancel/PEV/CNRS)

center. Each penguin spends an average of fifty minutes in the huddle, and most leave when they are again near the periphery, since birds at the center can hardly budge. Individuals leave the huddle when it has served its function for them and they feel excessively warm; if an entire huddle breaks up, there is sometimes a haze of warm air rising from the area, indicating a dissipation of heat. At any location, the huddle may last, with different birds entering and leaving, for several minutes to several hours. The number of huddles in a colony increases when air temperature and wind speed diminish and when solar radiation increases. In more extreme conditions, when both temperature and solar radiation decrease and wind speed increases, penguins form larger huddles, indicating that these are more effective in severe weather than smaller huddles (Ancel 2015).

In some places, birds are so few and the need for huddling to maintain body warmth through the night so great that different species join together in what sheltering locations they find. At 4000 m and higher in the Andes, twenty or more birds of different species may be found pressed against one another under rock ledges. These may include the Bar-winged Cinclodes (*Cinclodes fuscus*), the Rufous-collared Sparrow

(*Zonotrichia capensis*), the Bright-rumped Yellow-Finch (*Sicalis uropygialis*), and three species of sierra-finch (*Phrygilus*) (Dorst 1957).

IN TREES

For most landbirds that do not have access to cavities, the usual resort is branches of trees, especially if sheltered. Tree Swallows (*Tachycineta bicolor*), which in the winter will sometimes roost with several birds in a single cavity or birdhouse, also resort to huddling in exposed sites on cold days. During winter on coastal barrier islands, where the swallows feed primarily on bayberries, no trees may be large enough to have cavities. There, they huddle in small groups on low branches exposed to the sun and protected from wind (Elliott 1939). Golden-crowned Kinglets (*Regulus satrapa*), which never use cavities, always roost on branches. In Maine, where winter nights are sixteen hours long and temperatures may fall below −40°C, they have been found on small conifer branches in clusters of four, with the two outer birds facing inward, pressed against the bodies of the two central birds, all with feathers fluffed (Heinrich 2003).

In lowland Amazonia, groups of White-throated Toucans (*Ramphastos tucanus*) that spend the day foraging together roost side by side on a branch, with plumage fluffed, head tucked into back, and tail pressed down on the back, in the same posture as cavity-roosting toucans (Short and Horne 2001, p. 426). Birds that typically perch horizontally on branches, like nightjars and related families, can also huddle. In southern India and Sri Lanka, pairs of nocturnal Sri Lanka Frogmouths (*Batrachostomus moniliger*) rest by day in pairs, horizontally on forest branches, the birds tail to tail, facing in opposite directions but still in close body contact; sometimes a young bird rests in between them (Holyoak 2001, p. 147).

A migrating flock of a few hundred Vaux's Swifts (*Chaetura vauxi*), evidently not having found a more sheltered location such as their traditional hollow tree or more recently adopted chimneys, spent the night of May 7, 1964, tightly packed on the trunk of a tamarisk tree (*Tamarix* sp.) near the Arizona bank of the Colorado River south of the Davis Dam in Mojave County. The nighttime temperature was 3.3°C, relatively low for the season. The birds covered an area of the trunk 1.2 m long and .35 m wide and were often three birds deep, all facing upward, clinging to the trunk or another bird's back. All were torpid, and beneath the tree were a few dead swifts, but within two hours of sunrise all the surviving birds were on the wing (Stager 1965).

Configurations and Hierarchies

The configuration of huddlers, where this can be seen, indicates whether or how the birds position themselves to maximize the thermal benefits. The head is the part of the body that loses heat most rapidly, which is why most birds, when roosting alone, turn and bury it in the feathers of their back. When in a group, however, it may be more efficient to have the head facing forward, adjacent those of other birds. In Europe, the Treecreeper (*Certhia familiaris*) and the Short-toed Treecreeper (*C. brachydactyla*) have both been found in clusters of as many as fifteen on tree trunks, sometimes in two layers, clinging to the bark with all heads facing inward (Löhrl 1955). A group of thirteen Eastern Bluebirds (*Sialia sialis*) in a birdhouse in Indiana during a cold February night filled about three-quarters of the box; the entrance hole was blocked by the tail of the uppermost bird. All the birds rested on their belly, facing inward, with their heads pointed downward so that their bodies were at a 60° angle in two or three layers. This configuration may reduce the risk of suffocation, while the birds' exhalations may warm the air each bird breathes (Frazier and Nolan 1959).

Inca Doves (*Columbina inca*) in the Southwest huddle in pairs or small groups during winter nights, but in the daytime they form pyramids of two or three rows, with all birds facing downwind and into the sun, sometimes with as many as twelve birds in the group. Every five minutes or so, the outermost birds on the bottom row, which are receiving the least warmth from their neighbors, fly to the top, displacing the birds already there and precipitating a new arrangement. These pyramids last for about an hour, in morning or afternoon when the ambient temperature is less than −6°C, and on the same perches, on buildings as well as on branches, winter after winter (Robertson and Schnapf 1987).

In mousebirds, a family unique to Africa, each of the six species forms clusters of up to fourteen birds, not only during cold weather, rain, and at night but even during the hottest hours of the day. White-backed Mousebirds (*Colius colius*) in the deserts of South Africa are so sensitive to temperature that in early morning and late afternoon they sun themselves individually in trees; as soon as a cloud passes over the sun, however, they cluster together. Only a few birds in the group are actually perched on a branch; those on the periphery are clinging to other birds nearer the center (McKechnie and Lovegrove 2001).

Huddling is the norm among tropical cooperative breeding species, in which some young from prior years assist their parents with subsequent broods. In addition to the thermal benefits, body contact through the night may be a way the group reinforces social cohesion. Extended families of Green Woodhoopoe roost in cavities, but other cooperative breeding species settle on branches. In lowland Costa Rica, one adult male and four females or young male Purple-throated Fruitcrows (*Querula purpurata*) were found spending four nights pressed side by side on a thin branch of a small tree nine meters above the ground, lower than their usual foraging height in the forest (Snow 1982, p. 134). All cooperative-breeding cuckoos roost together, sometimes in piles. Guira Cuckoos (*Guira guira*) in South American dry savanna and scrub woodlands huddle in trees for the night and sometimes also during the day, with as many as fourteen together in a row; nevertheless, some die during cold nights. Smooth-billed Anis (*Crotophaga ani*) huddle not only at night, when their body temperature drops by 8°C, but also during rain (Payne 2005, pp. 135, 170, 176).

Australian Splendid Fairywrens (*Malurus splendens*) live in family groups with a dominant and a subordinate adult male, one or two females, and their recent young. They move through their shrubby habitat as a group. At the end of the day, they begin huddling first on one perch and then on another, with some jumping to get to the middle of the group, even as they are preening one another and chirping and trilling. As darkness falls, they move to a set roost in dense foliage where they huddle in a row, all facing the same direction with heads tucked into their chest and tails down. Before dawn, the adult males separate to sing, then rejoin the huddle, and at sunrise the group begins foraging (Schodde 1982, pp. 51–52).

Not all positions are equal within any group of huddling birds. Hierarchies develop when the huddle is more stable than the five-minute shifts in Inca Dove pyramids. Laboratory experiments with several flocks of communally nesting Long-tailed Tits (*Aegithalos caudatus*), in which each bird was individually marked and its rank recorded based on dominance interactions while feeding, found that dominant birds secured the inner places in the row of birds on a perch. There was always movement when the birds initially gathered on their perch, no matter whether the first birds there were dominant or subordinate within the flock. Those at the ends sought places in the middle while those already in the center of the row remained there. As the number of birds in the experimental roosts increased, there was less movement since more of the birds were surrounded by others. The most subordinate birds, however, were always at the ends of the row. Once the row had settled, the birds did not shift again during the night (McGowan et al. 2006). Similar experiments with Bronze Mannikins (*Spermestes cucullata*) showed that dominant birds had central positions in groups of five birds roosting in a compressed row in a cage. Dominance was similarly determined by the interactions of the birds at a feeder; the dominant individuals were able to consume more food than subordinates, so they began each night both better fed and more insulated than the others (Calf et al. 2002).

Among cooperative breeders in which families roost together and huddle, the hierarchy and configuration within the group varies among species, both for those that shelter in dormitories they have constructed and for those that sleep in a row on a branch. Australian Chestnut-crowned Babblers (*Pomatostomus ruficeps*), in the family Pomatostomidae of the Australo-Papuan region, live in extended families with female immigrants. They roost in bulky, domed, stick dormitories in trees, with an average of 8.5 birds, but sometimes as many as twenty-one. The group takes approximately six minutes to enter through a narrow opening; in the morning, it exits in the reverse sequence. In the evening, entry typically begins with dominant females and subordinate males belonging to the family, followed by the juveniles escorted by dominant males, and lastly, the immigrant females, not related to the core family. These females' position closest to the entrance leaves them most vulnerable to predation as well as benefiting least from the huddle, while the juveniles, to which the family members are all related, are in the center (Nomano et al. 2021).

The configuration is different in cooperative breeding babblers in the unrelated Afro-Eurasian family Timaliidae, which do not roost within nests. In Israel, families of Arabian Babblers (*Turdoides squamiceps*) huddle in a row on a branch of a dense thorn tree; the eldest male is always at one end and the other end is occupied by the second to fourth eldest male, usually the second. These individuals are most vulnerable to predation and have been seen captured and devoured by nocturnal snakes, but they may occupy the more dangerous positions because they are better experienced or because this is a form of social advertising that determines or demonstrates their status within the group (Bishop and Groves 1991).

Similarly, in a well-studied and individually marked extended family of Jungle Babblers (*T. striatus*) roosting in a row on branches in thorn scrub near New Delhi, the breeding male was on the inner edge, the breeding female beside it, then a second-year male, two first-year birds, a second-year female, and on the outer edge a nonbreeding adult male. In another family with younger birds, the juveniles, between two and six weeks old, joined the roost last by squeezing into the center of the group (Gaston 1977). In these species, the individuals most likely in greatest need of insulation are most surrounded by older birds; the male deepest into the tree may be safest from predators approaching from its edge, but evidently not from snakes that will climb up the trunk.

Benefits and Risks

Given how many birds at high latitudes only resort to communal roosting and huddling during periods of unusual cold, it is likely that, for these species, thermal benefits are the principal reason for this behavior, especially among birds that do not deploy hypothermia. Among highly social birds at lower latitudes that always huddle, whether cooperative breeders or not, it may be that this behavior is essential to their survival.

Common Bushtits (*Psaltriparus minimus*) are not cooperative breeders, but they live and roost in flocks of as many as seventy or more that all spend the night in the same dense foliage. Experiments showed that in an environment with a constant 20°C, bushtits consumed 80 percent of their body mass each day, or they lost weight if less food was available; at lower temperatures, they may need to consume 100 percent of their weight. At 20°C, even two birds roosting pressed against one another reduced the nocturnal energy cost of each by more than 30 percent, so the benefits of larger aggregations must be greater still. In the massive nest structures of communal—but not cooperative—breeding Sociable Weavers (*Philetairus socius*) in the Kalahari Desert, birds roost together in nest chambers during winter. A chamber with two birds is 5°C higher than the ambient temperature, while a chamber with a single bird is only 2.5°C warmer than an empty one (White et al. 1975). Finally, cooperative breeding Green Woodhoopoes kept at 20°C and below saved 12–29 percent of their daily energy expenditure when huddling, compared with birds roosting singly (Boix-Hinzen and Lovegrove 1998).

When huddling takes place in cavities, safety is another benefit, or at least a valued byproduct, especially in regions and ecosystems where the major nocturnal predators are owls, which capture birds in exposed situations but do not reach into cavities. Birds in ecosystems that include predators such as snakes and mammals that search and penetrate cavities, however, may sacrifice a degree of safety; the thermal benefits of a cavity, from which they cannot easily escape, must exceed what can be achieved when huddling in an exposed situation. Green Woodhoopoes go to great lengths to locate and secure roost cavities. If no suitable cavity is available within their territory, they will fly two kilometers across treeless grassland to reach an isolated roost tree, returning in the morning. They never huddle in thick thorny vegetation, where they could more likely escape from their several mammalian nocturnal predators; even in this vegetation, huddling would evidently not generate the warmth the birds require (Ligon and Ligon 1978).

In Australia, family groups of cooperative breeding Kookaburras (*Dacelo gigas*) regularly roost in two or three favored trees, gathering on the outer portion of a heavy branch 12–15 m off the ground. There, they press against one another, forming a single line rather than presenting the discernable silhouette of any bird. They do this at all seasons, including the hot Australian summer, so, especially for such a large-bodied bird, huddling seems not likely for thermoregulation. The assemblage may be an antipredator device focused on Powerful Owls (*Ninox strenua*), which prey on Kookaburras; together, the roosting birds may appear to be something too large to attack, while a single bird on an outer branch would be an easy target (Parry 1972, p. 56).

With the benefits of huddling can come several risks. The densely packed proximity for long hours, when birds may not be able to shift their place and which may be repeated for a season or a lifetime, facilitates the spread of pathogens and parasites. Skutch (1989a, p. 161) summarized many of the risks for birds in cavities: "The expedient of clustering tightly in closed spaces is not without special dangers. Birds deep in the feathered mass may suffocate; those on the outside may squander the last remnant of their depleted sources of energy by trying to push inward where they would be warmer; the plumage of some may be soiled by the excreta of their companions, reducing its capacity to retain heat; survivors near the bottom of the heap, lacking strength to push upward, may find themselves buried alive by the corpses above them. Clustering may be the last forlorn hope of birds dying of cold and hunger."

From cavities occupied by Pygmy Nuthatches, which roost in larger assemblages than are known for any other bird (at least one hundred), there are several accounts of dead desiccated birds found at the bottom (Knorr 1957). The remains of nuthatches in cavities suggest that those at the bottom were the first to suffocate, but during a period of severe spring weather in Manitoba, when ten Tree Swallows were sheltering in various cavities, eight dead ones were found in an old Downy Woodpecker (*Picoides pubescens*) nest, with the only two surviving ones at the very bottom. Both of these were very weak, and one died soon afterward; the other eventually recovered. But death from smothering, here probably exacerbated by starvation, is not restricted to cavities from which there is no easy exit. During the same Manitoba cold spell, Barn Swallows (*Hirundo rustica*) clustered in nests from prior years; at least eight were observed in one nest, in two layers, with the upper birds trying to penetrate deeper into the cluster. The next morning, one dead swallow was in the nest, and other dead ones were in nests elsewhere or on the ground below (Weatherhead et al. 1985). Whether more would have survived clustering in only a single layer is not known.

Still other hazards, with consequences beyond the immediate victims, have been reported from communal roosts of Acorn Woodpeckers (*Melanerpes formicivorus*), a cooperative breeder in which members of the group do not consistently roost together, although sometimes as many as fourteen have used a single cavity. Birds occasionally get stuck entering or exiting the cavity. In one case, two woodpeckers evidently trying to exit at the same time were stuck at their shoulders at the entrance and died there. Behind them, within the cavity, was another bird that had died for lack of any means of escape. This occurred in a colony where all the birds were known; the two at the entrance were breeding males, and the victim behind them was the group's breeding female. In their absence, the remaining subordinate members of the group could not defend their territory and were supplanted by another family (Stanback 1998).

Summary

Even for birds that do not roost in flocks at any time during their annual cycle, roosting behavior has social aspects. Among many territorial birds that live in pairs, the pair roosts together or close by during the breeding season, likely an effort by the male to prevent his mate from any extra-pair copulations. Nonmigratory pairs may spend the entire year roosting together for the same reason as well as to defend their territory and its best roost sites. For nonmigratory but less territorial species that shift to communal roosts in the nonbreeding season, these gatherings are a good place for young birds and single adults to find a prospective mate.

Young birds just out of the nest are at the most vulnerable stage of their lives, perhaps never more so than during the night. Most, however, receive no guidance from their parents about roost sites after the parents cease caring for them. Geese, cranes, and other birds that migrate as a family continue roosting together, with the young birds gaining from their experienced parents. A few sedentary birds are known to lead their young to a roost site. Nearly all of these species nest and roost in cavities or domed nests, or they build dormitories for use after the breeding season. Some parents then sleep with the young; others roost elsewhere nearby. Among long-lived birds like vultures, where young take several years to reach adulthood, family members roost near each other, the better the next day to search for carcasses from adjacent vantage points in the sky and then to share them on the ground.

Huddling together for warmth likely originated in the nest, where the body heat of young birds pressed together or incubated by a parent kept them warmer than each bird would be on its own. The habit persists among a wide variety of birds, from penguins in Antarctica to tropical toucans and others that may be particularly sensitive to cold, and it may take place during the day as well as when birds are roosting to sleep. In winter, unrelated individuals and, in extreme cases, unrelated species, may huddle together in sheltered spots. Small birds that use cavities for nesting will seek these out. Others gather in well-insulated unused nests of other species, in their own dormitories, in dense foliage, or, in places with no shelter, creating dense masses.

Huddling birds in a cavity or on a branch adopt the position that best reduces heat loss and receives the most benefit from adjacent birds. With this, however, come risks, including potential suffocation in a cavity and the spread of pathogens and parasites. There is also often a hierarchy among the huddlers, with dominant birds securing the best position, usually at the center of the huddle. Among cooperative breeders in which fledged young remain with their parents, this hierarchy can be more complex; in some, dominants are at the center of a huddle and subordinates outermost, but in others that roost in a row on a branch, the dominant males are at the extremities. This may be because they are most experienced and likely most alert during the night to any approaching danger.

6

Flock Roosts

The evening gatherings of hundreds, thousands, sometimes even millions of birds to roost are among the most dramatic spectacles provided by birds. Trails through the sky of birds approaching from different directions may begin kilometers away, guiding both other birds as well as observers to the site. There, birds may be flying overhead in excited noisy circles or may plunge directly and silently to find a preferred perch. These assemblages can be encouraging—a reminder that some birds are still amazingly abundant—or they are a useful quantifiable indicator that their populations have declined. Given that these birds have traveled long distances to reach the roost, many are sufficiently flexible that they can shift sites if a traditional one is destroyed. In some places, however, there may be few options of equal quality. Some spectacular communal roosts, such as the mangroves of the Coroni Swamp in Trinidad, where the island's and even some commuting Venezuelan Scarlet Ibises (*Eudocimus ruber*) gather in thousands, have become tourist destinations that generate revenue and justifications for conservation and protection.

The birds that roost year-round in large flocks often have different ecologies from those that roost solitarily, in families, or in relatively small aggregations. Some other species shift their roosting habits, just as they do their diet, from one season or locale to another. Communal roosting among large numbers of unrelated birds is most widespread after the breeding season, when many birds, even essentially sedentary ones, are no longer strictly territorial. Migrants such as shorebirds, swallows, and other birds that move about in flocks from the time they leave their breeding territory until they return the following year may resort, as they travel, to traditional roosts or settle in a new one each evening. There are other migrants, however, that, whether en route or at their wintering destination, spend the day foraging solitarily, even maintaining a daytime territory, but they also gather in flocks for the night. Across the spectrum of various species' annual and life cycles, there are some common denominators: individual birds most likely to roost in flocks are sexually immature and unpaired and do not hold a territory—at least one that requires constant presence—or they are migrating and wintering at a distance from their breeding grounds.

Within some avian lineages, flock roosting is widespread, in others it is rare. Small, territorial birds that roost in cavities, nests, dormitories, or concealing vegetation, and those that save energy at night by hypothermia or huddling in family groups, are not likely to evolve flock roosting habits. For many small birds, the energetic expense of a distant commute is also unsustainable. Among some families in which flock roosting is thought to be the ancestral behavior, it has been lost by some species. A survey of the origins of communal roosting that considered species in both large flocks and smaller groups concluded that, where

the habit has been lost, it is generally due to one or more of three ecological factors: a shift from flocking to solitary foraging, a shift from
nonterritorial to territorial behavior, or a shift from nocturnal to diurnal
roosting (Beauchamp 1999).

Benefits

There are several benefits of roosting in large aggregations, and birds
may enjoy more than one benefit simultaneously. Lower thermoregulatory costs can come from the presence of many other birds nearby, by
blocking wind or generating heat even when their bodies are not in
contact. Predation risk is likewise lowered by the presence of many
other birds, since at any moment some are likely to be awake and vigilant, and the odds of being the one caught are reduced—with the least
risk for those far from any roost's periphery. Communal roosts may also
increase foraging efficiency by bringing together more birds that the
next day can search for food, especially if it is patchily located, where
solitary birds might have less chance of finding it. In addition, birds
returning the next morning to a place where they will resume feeding
may be followed by others at the roost that need to find a new foraging
site; these additional birds may be useful if they increase the number
present that are vigilant (Beauchamp 1999).

A review of communal roosting in Indian birds (Gadgil and Ali 1975)
noted some ecological correlates with large roosts and sought to determine whether these were linked to any benefits. Most of the Indian
species that gather for the night in large flocks live in grasslands, scrub-
lands, and cultivated habitats, and they feed on the ground. Few dwell in
woodlands, and these birds roost communally only seasonally and not
with other species. Most flock roosters also feed in flocks; few birds
that forage alone roost communally, and still fewer among the species that
spend the day moving as pairs. Thirty-three of the fifty-nine Indian species known to roost communally never roost with other species; their
roosts are usually far smaller than those of the species that form very
large aggregations. The six species with the largest roost populations,
in at least the thousands, all have multispecies roosts. Some of these

contain birds using similar feeding habitats, such as three wagtail species that roost together, or the Rose-colored Starling (*Sturnus roseus*) and Common Myna (*Acridotheres tristis*), while others are entirely dissimilar— such as Cattle Egrets (*Bubulcus ibis*), Rose-ringed Parakeets (*Psittacula krameri*), and House Crows (*Corvus splendens*). The authors conclude that in India avoiding heat loss is not a motivation for communal roosts, and the opportunity to learn of food sources by following roost-mates is not relevant to the few species that disperse from roosts to feed solitarily. Nor can species in most mixed roosts learn much of use from the other species sharing them when these have very different ecologies. Reducing each individual's risk of predation seems the most likely benefit, even as the conspicuousness and long-term use of communal roosts must make these well known to predators.

At a mixed-species roost, some may gain more benefits than others. In Puerto Rico, diurnally territorial Gray Kingbirds (*Tyrannus dominicensis*) fly singly or in pairs to offshore mangrove cays, some as far as 5.5 km from their territories. Flocks numbering up to 150 spend the night, sometimes sharing the cays with thousands of Cattle Egrets and various blackbirds. The cays are only 30–35 m from the mainland, so they are easily accessible to the owls known to prey on roosting birds in Puerto Rico; more distant cays are not used. Kingbirds gain no information about relevant food sources from egrets or blackbirds before they return each morning to their own territory, so the reasons for their travel are not clear, especially since during the breeding season many Gray Kingbirds roost within their territory. The kingbirds arrive later and depart earlier than their egret and blackbird neighbors, suggesting that at least some must travel farther from their territory than the other species, or that they reduce their exposure to hawks when light is faint (Post 1982). The benefits to kingbirds are not obvious, but at some of these Puerto Rican cay roosts wintering Northern Waterthrushes (*Parkesia noveboracensis*), which similarly each maintain a diurnal territory, commute as many as two kilometers to reach. Some waterthrushes form groups of up to twelve birds, each bird within one to three meters of others; these groups are at the periphery of kingbird roosts. The waterthrushes may seek out the kingbirds, which are more alert to disturbance (Smith et al. 2008).

For some species, the benefits of roosting in large flocks are even less obvious. Eagles, hawks, and falcons, which sometimes gather in hundreds or more, are not at risk of predation and are too large and perch too far apart to gain any thermal benefits. They hunt solitarily—unlike scavenging vultures—with no intent to share their prey, and the roost sites that attract them are not in short supply. For smaller birds that during the day gain some protection from predators by feeding in a flock, it has been hypothesized that this benefit peaks at less than one thousand birds in a group, based on the frequency with which each individual in the group stops feeding to scan for possible danger. When that frequency is spread out among a thousand birds, the chance that at any given second no bird in the flock is vigilant is close to zero (Pulliam 1973). While the scanning rate is likely less when birds are sleeping than feeding, the number of birds necessary in a roost to maintain a high frequency is still far lower than the orders of magnitude found in many such aggregations.

Single-Species Roosts

Flock Roosts on Migration

Migratory birds that no longer need a territory when they have completed their breeding cycle may then change their roosting behavior. Some territorial birds even form roosts long before they have completed nesting. Male Scissor-tailed Flycatchers (*Tyrannus forficatus*) roost in flocks while their mate is incubating; the males return to their territories every day to defend it and, in due course, to feed their young. Any nonbreeding females or failed breeders also join the roost relatively early. When the young fledge, they are led to the roost. In Texas, from August, when nesting is completed, into early October, when the flycatchers depart, they continue roosting together, sometimes with more than two hundred in a single tree; some birds have traveled more than two kilometers to reach the roost (Skutch 1997, pp. 36–37). Similarly, when their young fledge, Purple Martins (*Progne subis*) gather in large premigratory roosts. For them also, the first to assemble are probably failed breeders or young nonbreeders, then joined by adults and their mobile young, followed by some migrants. In South Carolina, at least 700,000 martins roost on the 5-hectare Lunch Island, in a lake near Columbia, starting late June, with numbers peaking in late July, and continuing until late August, when most have departed. Every morning, the birds disperse in all directions, some going as far as 100 km from the roost during the day (Russell and Gautheaux 1999).

Like many shorebirds that breed scattered over the Arctic tundra, Dunlins (*Calidris alpina*) afterward move to congregate at estuaries where they spend the day and, depending on the tides, sometimes part of the night, feeding intensely during their molt and before undertaking the first of their long flights. On the Yukon-Kuskokwim Delta, Alaska, 70,000–100,000 Dunlins feed at Angyoyaravak Bay and roost during diurnal high tide in flocks of up to 21,500, usually on mudflats. At some sites, birds come and go, with some birds staying only a few minutes before moving to another roost, even as most birds around them sleep (Handel and Gill 1992). The other species of small sandpipers that are also abundant in the bay seem each to have their own roosts. This may reflect ecological preferences or responses to predation, since when sandpipers typically take off in dense flocks it is likely more efficient if all in the flock are the same size and move at the same speed. Where large numbers of Bar-tailed Godwits (*Limosa lapponica*) and Dunlins roost on mudflats in Alaska, they may be within tens of meters of each other but in separate flocks (McCaffery and Gill 2001).

Spring migration is also a time for flock roosts. In southern Quebec, near Baie-du-Febvre, as many as 500,000 Snow Geese (*Chen caerulescens*) use a 5-sq km area of flooded lowlands to roost in spring—while feeding sometimes as far as 62 km away—before their final long flight to their arctic breeding grounds. Individual birds stay there two to thirty-two days. While access to abundant agricultural food sources and safety from predators in wet areas may seem the reasons for these site choices and aggregations, the geese are most faithful to areas near woodlands and a main road rather than areas with less disturbance or where predators would be more conspicuous (Béchet et al. 2010). Similarly, on spring migration, 300,000 Sandhill Cranes (*Grus canadensis*)

Flocks of migrating Sandhill Cranes gather for the night in the shallows of rivers, like the Platte in Nebraska, where they are safer from predators.

may roost together in the shallows of the Platte River in Nebraska, remaining there for days or weeks and dispersing over a wide area of farm fields each day to feed (Sparling and Krau 1994). Cranes and geese are among the few species that migrate as family units, in autumn and to a somewhat lesser extent in spring; as a result, young birds learn the roost sites favored by their parents and carry the knowledge from generation to generation.

The raptors migrating between North and South America through the narrow parts of Mexico and Central America are less particular about their roost sites but, by virtue of their numbers, they form very large aggregations during the few weeks in autumn and spring when they pass through this region. Skutch (1989a, p. 28) describes a northbound flock of thousands of Swainson's Hawks (*Buteo swainsoni*) descending to a pasture on the Pacific slope of Costa Rica in late afternoon, when the thermals carrying them dissipate. Most of the birds found a perch in the few trees dotting the pasture and surrounding it, while others settled for the night on logs, stumps, and the ground itself. Safety seems not a concern, and proximity to food is definitely not one, because

Swainson's Hawks resume their flight every morning, usually without hunting. Unlike the cranes and geese that fly by continuously flapping, and remain for days at a migratory staging area to fatten up for the next long lap, Swainson's Hawks riding thermals require so little energy that they are able to fast for days as they soar over Mexico and Central America, until they can spread out through a larger landmass and hunt without competition. Fasting, however, requires that each bird not tarry, so no roost site is used by any individual for more than a night unless rain prevents the sun from creating thermals.

At their winter destination, flock roosters may move among nearby roosts. In southeastern Louisiana, more than one million Tree Swallows (*Tachycineta bicolor*) roost in several aggregations in fields of sugar cane covering 120 square km, dispersing to feed sometimes tens of kilometers away each day. The same fields were used most nights; whenever a flock left one field, it was to use another within 2–3 km. Individually tracked birds were found to return to the roost they had used the previous night 60 percent of the time, and when they moved to another roost, it was usually within 15 km of their prior one, well within the range a swallow might travel over the course of the day while foraging (Laughlin 2014).

Flock Roosts of Sedentary Species

Many nonmigratory birds also form single-species roosts. These may be used during many months of the year and may draw birds from an extensive breeding area. Some sites are traditional, others short-term. On Staten Island, New York, a winter roost near a landfill attracts ten thousand to fifteen thousand American Crows (*Corvus brachyrhynchos*) that, for the most part, return each morning to their territories, held by cooperative groups, several kilometers away. During spring, individually marked crows were found to roost 87 pecent of the time within their territory, but in winter, when more birds spend more time feeding beyond their territory, only 42 percent of the crows roosted in it (Caccamise et al. 1997). In California, communal roosts of Yellow-billed Magpies (*Pica nuttalli*) that are used outside the breeding season may include several hundred birds from different breeding colonies; the roosts are away from any of the colonies (Birkhead 1991, p. 81).

For most corvids that roost in flocks, these are of single species. Few other birds seem ever to roost with corvids, which, while their size and aggressiveness may deter predators, could in turn attack any smaller birds nearby. At Malheur Lake, Oregon, Common Ravens (*Corvus corax*) used the same marsh roost site for at least ten to fifteen years prior to a study in 1975 that counted as many as 836 on January evenings (Stiehl 1981). A still larger raven roost, with approximately 1,500 birds, existed, however, for only thirty-eight days (November 8 to December 15, 1987) in the Mojave Desert in the foothills of the San Gabriel Mountains of California; there, as in the Oregon marsh, the birds used vegetation less than 3 m high. The birds approached the site individually and in small groups from several directions, and they departed similarly in the morning. None had been seen there for at least nine years prior, and after every bird left on December 15, no comparable aggregations were seen over the next five years (Cotterman and Heinrich 1993).

In parts of the world with several species of closely related corvids that roost near each other, most maintain a degree of distance within shared roosts. At a tree plantation roost in Cornwall that held up to 150 ravens, about 200 Carrion Crows (*C. corone*), 2,500 Rooks (*C. frugilegus*), and 7,000–8,000 Jackdaws (*C. monedula*), the ravens and crows each used a separate area, while the Rooks and Jackdaws, perhaps because of their overwhelming numbers, settled in more or less the same section of the plantation. At another Cornish roost, Rooks settled in taller trees, Jackdaws in shorter ones, and Carrion Crows used a part of the plantation more than 90 m away. At a third roost of only Rooks and Jackdaws, both used tree tops, but each species in trees in a different part of the roost (Coombs 1978, pp. 66, 104, 125).

When birds from large communal roosts that are used much of the year disperse each day to feed, they have different degrees of fidelity to their previous roost. Nightly turnover at Black Vulture (*Coragyps atratus*) roosts in Florida is 33 percent (Stolen and Taylor). Jackdaws from different roosts disperse to feed over a wide area, where they encounter Jackdaws from other roosts. In the afternoon, when they are returning to their roosts, flight lines of birds headed in different directions may cross (Coombs 1978, p. 124). The fidelity of individual Jackdaws to a roost has not been studied, but since they usually spend the day in pairs and roost side by side, the decision to change roost sites is likely to be a joint one.

Multispecies Roosts

Multispecies roosts may include species closely related taxonomically and overlapping ecologically or those whose diurnal lives are entirely different. Each of the various species may be there year-round or only seasonally in greater or lesser numbers. As with single-species roosts, use of the roost may vary by age class, sex, and breeding status of individuals of each species.

Some of the largest multispecies roosts are found in the southern United States. These roosts are used throughout the year by local birds, which are then vastly outnumbered as migrants from farther north arrive for the winter. Then, several million icterids, a mix of Red-winged Blackbirds (*Agelaius phoeniceus*), Common Grackles (*Quiscalus quiscula*), Brown-headed Cowbirds (*Molothrus ater*), Rusty Blackbirds (*Euphagus carolinus*), and European Starlings (*Sturnus vulgaris*) use extensive wetlands, rice fields, canebrakes, deciduous thickets in swamps, or stands of conifers. American Robins (*Turdus migratorius*) may join, but they stay at the periphery. Over the course of the winter, the populations and sex ratios of each species using the roost may change (Meanley 1965).

In much smaller numbers, birds of prey also form multispecies roosts. In Nevada, a grove of cottonwoods approximately 150 by 100 m was used in winter by 180–210 buteos, including about ten Rough-legged Hawks (*Buteo lagopus*), thirty-five Ferruginous Hawks (*B. regalis*), and the remainder Red-tailed Hawks (*B. jamaicensis*), plus three Great Horned Owls (*Bubo virginianus*) (Bechard and Swem 2002). On the Isle of Man, a stand of conifers 20 m high, similar to many nearby sites equally available, attracts between late August and early December 5–9 Peregrine Falcons (*Falco peregrinus*), up to five Common Kestrels (*F. tinnunculus*), two Merlins (*F. columbarius*), four to six Eurasian Sparrowhawks

(*Accipiter nisus*), and as many as seventeen Common Ravens. The smaller species approach the woodland fast and low, shooting up into the branches as they reach the trees, while the Peregrines circle high over the site, often for twenty minutes. The ravens fly in just above the treetops and ignore and are ignored by the Peregrines. While the smaller raptors might in other situations be prey for the Peregrines, here the density of the conifers prevents attacks. Carrion Crows, which elsewhere harass these species, avoid the roost, while in open areas on the island both they and gulls mob Peregrines. So, peace and quiet may be the significant benefit of this multispecies roost (Kelly and Thorpe 1993).

For shorebirds in the nonbreeding season, high tides, whatever the hour, limit the space available for roosting and may force the mixing of species that otherwise separate by feeding habitat or method. On Laysan Island in the Pacific, flocks of Wandering Tattlers (*Heteroscelus incanus*) rest dozens of meters away from mixed-species roosts of Pacific Golden-Plovers (*Pluvialis fulva*) and Ruddy Turnstones (*Arenaria interpres*), but on the coast of California tattlers join the assemblages that may contain Black Oystercatchers (*Haematopus bachmani*), Hudsonian Whimbrels (*Numenius hudsonicus*), Willets (*Catoptrophorus semipalmatus*), Surfbirds (*Aphriza virgata*), Rock Sandpipers (*Calidris ptilocnemis*), and Black Turnstones (*Arenaria melanocephala*) (Gill et al. 2002).

Migrating and wintering birds may often be attracted to roosting sites used by residents. In Costa Rica, migrant Scissor-tailed Flycatchers roost in trees with local Tropical Kingbirds (*Tyrannus melancholicus*) (Skutch 1987, p. 37). Sometimes the roosting sites differ from the migrants' usual choice. In a patch of tall introduced elephant grass (*Pennisetum*) in a cleared valley in Honduras used principally by White-collared Seedeaters (*Sporophila torqueola*) and three related species of similar ecology, they are joined in August by migrant Orchard Orioles (*Icterus spurius*), then by Eastern Kingbirds (*T. tyrannus*), and in October by Baltimore Orioles (*I. galbula*), all of which more frequently roost in trees. The migrants may be gaining protection from predators by sharing this roost (Skutch 1989a, p. 51).

In the high veld at 1600 m in the Transvaal of South Africa, the dense reedbeds of marshes around permanent and impermanent lakes are a roosting magnet for birds of all habits in this otherwise dry and open landscape. At one such marsh 500–600 m in diameter where roosting habits were studied, the resident birds using it included Hadada Ibis (*Bostrychia hagedash*), Cattle Egrets, Spur-winged Geese (*Plectropterus gambensis*), Rock Pigeons (*Columba guinea*), Pied Starlings (*Spreo bicolor*), Red-billed Queleas (*Quelea quelea*), Golden Bishops (*Euplectes afer*), and Long-tailed Widows (*E. progne*). These were joined by vast numbers of wintering Barn Swallows (*Hirundo rustica*), but none of the dozen widespread resident swallows of the Transvaal, although watched for, were observed (Rudebeck 1955).

A particularly complex mixed-species roost has been observed in Zimbabwe, where birds of many ecologies regularly come together to use reedbeds that have willows at their edges, thereby providing both dense shelter and elevated perches. The early arrivals include Red-winged Starlings (*Onychognathus morio*), Southern Masked Weavers (*Ploceus velatus*), Red-billed Queleas, Red-collared Widows (*Euplectes ardens*), Red Bishops (*E. orix*), and other finches, plus, during many months, wintering Yellow Wagtails (*Motacilla flava*) from Eurasia. Smaller numbers of Black-shouldered Kites (*Elanus caeruleus*) come in, mostly after sunset, flying low singly or in pairs and dropping into the reeds. The finches show no fear of the kites, but they tend to occupy different parts of the reedbed. Golden Weavers (*Ploceus xanthops*) continue to nest in reedbeds when both kites and smaller birds are roosting around them. The kites, in turn, leave areas where flocks of starlings are still noisy, returning when the last of these have arrived and settled. In the adjacent willows are Hamerkops (*Scopus umbretta*), Southern White-crowned Shrikes (*Eurocephalus anguitimens*), and Gray Herons (*Ardea cinerea*) (Brooke 1965).

Hierarchies

At large communal roosts, both single-species and multispecies, there is often a hierarchy based on age, gender, and condition, with dominant birds securing and retaining the more favored places. Some places are superior because they are closer to feeding areas or provide more thermal benefits, ease of landing and takeoff, greater visibility for spotting predators, or avoidance of droppings coming from neighboring birds.

Single-Species Hierarchies

Among the thousands of Snow Geese at Baie-du-Febvre, where birds show greater fidelity to one part of the roost over another, it is dominant birds that return there most often, and these birds arrive earliest each afternoon. By securing a place in the roost that had the lowest travel cost to feeding areas, the dominant birds probably maintain the superior condition that reinforces their dominance (Béchet 2010).

For Eurasian Oystercatchers (*Haematopus ostralegus*) as well, proximity to feeding areas also correlates with hierarchies at roost sites. On Texel and Vlieland, two islands on the Dutch Wadden Sea, occupied respectively in winter by forty thousand and twenty-five thousand oystercatchers, there are several roosts of different sizes. The largest roosts are nearest to the most expansive and best feeding areas, where the topography exposes flats for more hours each day. Adult female oystercatchers, larger than males, are more populous at these roosts, while midsized roosts have more subadults, and the smallest roosts have the most males and birds in their first year. Individuals with external tumors, missing toes, or malformed beaks are commonest at the smaller roosts, no matter what their age, and these roosts have the highest mortality (Swennen 1984).

Among some other shorebirds that roost in smaller flocks, typically of hundreds, the configurations are different. In Alaska, male and female Bar-tailed Godwits roost together, but juveniles are usually at the periphery or in separate flocks (McCaffery and Gill 2001). Adult Redshanks (*Tringa totanus*) wintering in Great Britain force juveniles to the windward side of mudflat roosts, where they lose whatever microclimate benefits there may be in joining a flock. For these young birds, the benefits must be in predator detection and avoidance (Ydenberg and Prins 1984). Dunlins, in contrast, have more juveniles in the center of roosting groups in Bodega Bay, California, and more of the birds in the center are females, which are larger than males (Ruiz et al. 1989).

Landbirds roosting in wetlands compete for the most favored sites, with those of the dominant sex, age, or size securing the places most sheltered from predators and weather. A study of male Red-winged Blackbirds roosting in a cattail stand at a pond in Ontario found that adult males settled in the densest, most central part of the marsh and, within that, preferred the areas where water was deeper. They often chased males in immature plumage, while the immature birds rarely chased others. The adult males gained the benefit of more protection from wind and perhaps also from terrestrial predators where the cattails were in deeper water, while the peripheral birds, at the edge of the pond, while also over deeper water, were more exposed to wind (Weatherhead and Hoysak 1984).

On a far smaller scale, in southwestern Scotland, as many as thirty Northern Harriers (*Circus cyaneus*) may roost together in a wetland. The females, larger than males, drive out adult males, and adults of both sexes may drive away first- or second-year males. Dominant birds seeking an occupied spot dive with lowered feet at the bird on the ground. The most persecuted individuals fly back and forth over the roost area, only finding an undisturbed spot after the others have settled (Watson 1977, p. 246). Similarly, Snail Kites (*Rostrhamus sociabilis*) in Florida roost communally in wetlands and adjacent woody vegetation, the birds usually 0.3–6.5 m apart, with adult females dominant over males and immatures (Sykes et al. 1995).

Species using trees may have a three-dimensional hierarchy. Among Rooks in East Lothian, Scotland, adults at a roost take the perches in the tops of trees and younger birds use lower branches. Since these winter roosts are often in leafless trees, where birds in the canopy are exposed to more wind, the advantages to adults are not obvious, especially since they have no nocturnal predators that might approach from the ground. They may be in sufficiently good condition that most winter nights are not a thermal challenge, while the lower birds—typically younger and less well-fed—benefit from reduced heat loss. When the weather is severe, however, the adults shift to these lower limbs, forcing the young birds off their perches and lower still, sometimes to where there are no branches. These birds must then leave the tree, sometimes in the dark, to search for another spot, where they may be more exposed. Most of the birds benefit most of the time from their place in the roost, but in trying conditions the most subordinate birds, in poorest condition, suffer most (Swingland 1977).

Cliffs likewise produce vertical hierarchies. On the Isle of Islay in the Inner Hebrides, 100–130 Red-billed Choughs (*Pyrrhocorax pyrrhocorax*) roost on a cliff protected from the prevailing westerly and southwesterly winds. Here, in an area of 80–100 sq m, most of the birds find spots to spend the night, favoring ledges, crevices, and small areas with vegetation. Third-year birds are highest, followed by second- and first-year birds at lower elevations, respectively. Third-year birds are most likely to attack others, with first-year birds attacked most often. These youngest choughs roost in lower and peripheral sites with the least density of birds. The predators of choughs at Islay are Peregrine Falcons and Common Buzzards (*Buteo buteo*); the choughs higher on the cliff may have an advantage in spotting an approaching raptor, as well as greater airspace for takeoff and evasion if the raptor is attacking from above (Still et al. 1987).

In the gigantic Swiss roosts of wintering Bramblings (*Fringilla montifringilla*), the individuals in the preferred sites have higher body mass, but not larger size, than birds on the periphery. Age and sex alone are not determining factors, but within each age and sex class, birds at the center of the roost are heavier than those at the periphery, indicating that they have been the most successful foragers that day. Adult males, however, are typically larger than females and, of the adults in the center part of one roost, 53 percent were male while on the periphery only 33 percent were male. Adults enter the conifer grove later in the day than juveniles and may then displace the young birds from the perches surrounded by the thickest vegetation (Jenni 1993).

In an English roost in leafless trees, flocks of 100,000 European Starlings show a similar pattern. The birds in the center are heavier than average, more often adult males, and the periphery is populated mainly by lighter and younger birds, with a higher proportion of females. At these roosts, it takes about twenty minutes following the arrival of the last birds for all individuals to find a place for the night. Fights are rare, but there is frequent aggressive posturing, which may suffice to signal relative rank (Summers et al. 1987).

Mixed-Species Hierarchies

At mixed-species roosts, the spatial configurations can be even more complex.

The several species in the southern US blackbird roosts each occupy a different place in the vegetation. At two Arkansas roosts in deciduous thickets, starlings roosted highest in the trees, then Common Grackles and male Red-winged Blackbirds. Below them were Brown-headed Cowbirds and female red-wings, and lower still were Rusty Blackbirds and more female red-wings. Whether this is by actual preference or due to the dominance of some species over others has not been tested. Birds at the highest points are more exposed to wind, rain, and aerial predators; the birds farther down are more concealed, but they would have a harder time escaping, and those closest to the ground are in easiest reach of terrestrial predators (Meanley 1965).

Avoiding the droppings of birds directly above may play a role in this stratification. The consequences from exposure to droppings are real. Uric acid and fecal matter have been shown to affect the water-repellant and insulating properties of feathers. In experiments with European Starlings, the individuals with lower roost positions that had accumulated droppings on their plumage were heavier after fifteen minutes of exposure to artificial rain than the clean birds from the uppermost perches. Some died within thirty minutes of the wetting (Yom-Tov 1979).

Some birds seek to avoid this hazard. The flocks of three species of Puerto Rican blackbirds that travel to offshore mangrove cays use the inner parts of the cays, roosting in the middle and lower levels of the canopy, where the foliage conceals them from potential predators and protects them from wind and rain. But with these advantages come the hazards of droppings from above. While the blackbirds may share their roost with Mourning Doves (*Zenaida macroura*) and Gray Kingbirds, they avoid the areas used by larger birds, including Cattle Egrets, Brown

Pelicans (*Pelecanus occidentalis*), and Magnificent Frigatebirds (*Fregata magnificens*), which all perch in the uppermost parts of the mangroves, where they can land and take off most easily (Post and Post 1987).

Roosts as Information Centers

Another benefit to some birds roosting in flocks may be the information they gain about food sources to seek out the following day. This hypothesis was first elaborated by Peter Ward and Amotz Zahavi (1973). They noted that many of the species that roost in flocks feed on food that is patchy in space or ephemeral in time but in such abundance that a large number can exploit it simultaneously without conflict. Individuals returning to a roost after a day of poor foraging can watch other birds depart the next morning to places where they fed the day before, just as in a breeding colony the poor foragers might note through the day which birds come back with a bulging crop or visible food for their young and follow them on their next feeding sortie.

In most cases where roosts may serve as information centers, there is no particular benefit received by the knowledgeable birds departing in the morning for a place where they expect to continue finding food. They are not actively or deliberately advertising their knowledge; the transfer of information is not intentional. Were these birds solitary feeders at an abundant source, they might gain from the presence of others if this resulted in increased vigilance and reduced predation risk, but most birds that roost in large aggregations also feed in flocks, so no individual is likely to have unique knowledge of a feeding site or be there alone.

While the information is most useful to conspecifics, at mixed roosts occupied by ecologically similar species, as at those of shorebirds, starlings and New World blackbirds, European corvids, or African finches and weavers, these birds may also be alert to the behavior of all their neighbors. Where roosts contain birds as ecologically disparate as Cattle Egrets and parrots, other features and benefits of the roost must be drawing them together, and the birds of each species are not likely giving attention to the departure patterns of the others.

A few other types of information are available at communal roosts that in fact have benefits both to signalers and receivers. In some of the mixed-species flocks that move together through tropical forests, the birds remain close when they settle for the night. At Barro Colorado Island, Panama, the nuclear species of understory mixed-species flocks are pairs or families of Checker-throated Antwrens (*Myrmotherula fulviventris*) and Dot-winged Antwrens (*Microrhopias quixensis*) that hold large, completely matching territories. They are joined by at least six other species that have territories within or overlapping these. The two nuclear species roost in the same tree, and some of the others do as well, or within fifteen meters. By staying close together, members of the flock know where to find one another, so they can set off early in the morning without any time lost searching for the nuclear species or being more vulnerable when moving about alone (Gradwohl and Greenberg 1980).

Other types of information that have been proposed as benefits for members of the same species when roosting together include the identification of potential mates and the knowledge of hierarchy among individuals that regularly roost together, the synchronization of molt and of group flight during migration, the learning of song dialects within regional populations, and the identification of novel predators (Bijleveld et al. 2010).

How Birds Watch and Follow Others at the Roost

FOR INFORMATION ON FEEDING SITES

Ward and Zahavi (1973) further suggested that the watching behavior of individuals that want to follow more knowledgeable birds varies by the topography of the roost site. When birds roost in treetops, on cliffs, or in other situations that provide a clear view of departing birds, the birds likely to fly first in the morning are those returning to where they fed the day before; birds in need of a new foraging site wait at the roost and watch where these others are going. In contrast, among the birds that roost in sites with little visibility, such as reedbeds, the less knowledgeable birds are likely to move to a spot with greater visibility, either before or after others depart, and from there watch and then follow the birds flying directly to feed.

At the roosts of Red-billed Queleas (*Quelea quelea*) in West Africa that sometimes contain millions, the birds leave in waves, with most flying straight off out of sight. Others, however, go only a few hundred meters, settling in trees or bushes, and join birds flying out in a subsequent wave (Ward 1965). Cattle Egrets in South Africa departing from a roost show a similar form of monitoring. The birds leave their roost in waves, but of two types: some birds set off purposefully on set flight lines, while others are indecisive. These may waver before pursuing one of the direct flight lines, or circle and join the next group flying in that manner, or they may return to the roost and depart later with other birds. This pattern is most frequent at the end of the local dry season, when the roost holds many more naive juveniles and when good feeding spots are both patchy and scarcer. In winter, when food and feeding places are abundant and widespread, most birds leave the roost in direct flight (Siegfried 1971).

BROOD PARASITES

Some birds that deposit their eggs in the nests of other species use communal roosts as information centers of a different sort. In Argentina, female Shiny Cowbirds (*Molothrus bonariensis*) search for nests of several species at the right stage of incubation in which they can deposit an egg the following day. Females roost communally while males, which are not directly involved in the parasitism, do not. Females preparing to lay an egg depart the roost before dawn, reaching their target nests when the incubating host female is most likely to be off it. Knowledgeable female Shiny Cowbirds are often trailed by another female that needs to lay an egg but has not found a nest. While the knowledgeable cowbird gets to the nest first and deposits her egg, her follower may then go to the nest, remove the first cowbird's egg, and replace it with her own (Scardamaglia et al. 2018).

In Puerto Rico, Shiny Cowbirds use actual nest hosts for information transfer at their joint roost. Yellow-shouldered Blackbirds (*Agelaius xanthomus*), Shiny Cowbirds, and Greater Antillean Grackles (*Quiscalus niger*) roost at predator-free sites including small islands, electric power stations, and isolated palm trees. About the same number of blackbirds and grackles use the roosts throughout the year, but during the blackbird's breeding season, in summer and autumn, cowbird numbers increase. In the morning, the cowbirds follow the blackbirds returning to their territories, thereby likely learning where these may have nests and be preparing to lay eggs. The parasitic cowbird reached Puerto Rico in the mid-twentieth century, in the course of its northward expansion through the Caribbean islands. Its parasitism of Yellow-shouldered Blackbirds is the principal cause of that species' recent rapid decline to endangered status (Post and Post 1987).

Testing the Hypothesis

The existence and effectiveness of roosts as information centers vary across species, seasons, and ecosystems. For roosting aggregations rather than breeding colonies, where there is traffic throughout the day, the hypothesis that naive birds follow knowledgeable ones to a food source requires that the food item lasts into the following day. This has been most effectively tested with species that consume items that, as long as these last, remain in the same place. Knowledgeable birds returning to a source they have discovered the day before will fly in a very specific direction rather than in a general scan over an area that was previously profitable, as might gannets searching over the sea for fish, or swallows over a marsh for hordes of emerging insects. The ideal food source for such tests is carrion, which will not shift location and which experimenters can supply or remove to examine different aspects of birds' responses. Thus, the most conclusive work has been done with vultures, ravens, and other scavengers.

VULTURES

In experiments with a known North Carolina population of Black Vultures (*Coragyps atratus*), including many individually marked birds, some that were the first to discover a baited carcass returned earliest the next day, with more birds from their roost. Other individuals from the population were held in captivity to ensure that they were ignorant of that carcass. When released two evenings later, they went directly to the roost; the next morning, these naive birds did not depart the roost until others had, and they then appeared at the carcass. It cannot be ruled out, however, that these naive birds, both experimental and control individuals, only reached the carcass after dispersing some distance from the roost and then watching others descend to it—known as "local enhancement." But Black Vulture roosts are largest during winter, when food is scarcest, because birds from smaller roosts then congregate at fewer sites, so the opportunities for information transfer are greater. While the knowledgeable birds may not be deliberately recruiting others, they benefit at the carcass when more birds are present to tear it apart. The greater number present also reduces any single bird's predation risk. Since adults are generally better than younger birds at finding carrion and are dominant at a carcass, the first finders typically gain from the presence of other vultures without any loss of access to their food (Rabenold 1987b).

At an Ontario roost of individually marked Turkey Vultures (*Cathartes aura*), the birds that found baited carcasses returned to them early the following morning—despite the lack of thermals that would have aided their flight—but they were not followed or immediately joined by any other birds from the roost. Other roost-mates that came to the carcass later were thought to reflect local enhancement rather than recruitment. This may reflect a difference between Turkey and Black Vulture ecology: Turkey Vultures prefer to scavenge much smaller carcasses, and dominant individuals do not tolerate others attempting to feed with them. The incentives for naive vultures at the roost to follow one that departs early may therefore be weak (Prior and Weatherhead 1991). For Turkey Vultures, and indeed sometimes for Black Vultures and other large scavenging birds, a benefit of the communal roost is that it brings together many birds that will then search adjacent areas that might not be within visual range if the birds roosted solitarily at places more spread apart (Buckley 1996).

RAVENS

Observations and experiments with Common Ravens have shown more complex and possibly more deliberate information sharing at roosts. In Maine, ravens feed primarily on carrion through the winter. While pairs of territorial adults have their own roost, nonbreeding and immature birds form communal roosts that—in the absence of a nearby steady source of food like a landfill—change, on average, every four nights, so that birds are within one kilometer of their current source. At the communal roosts, morning departures are highly synchronized, with most birds flying in the same direction. The morning after a carcass is found, the knowledgeable birds at the roost are the first to depart, returning to the carcass before sunrise. They are rapidly followed by others that, it seems, do not want to lose sight of what they take to be their guides. On subsequent days, when the birds know the location, some do not arrive so early or in such large groups (Heinrich 1994).

Ravens experimentally removed from a roost—to ensure that they were unaware of a newly placed carcass—and then returned to the roost the next day followed roost-mates the next morning, just as do Black Vultures. At the same time, individuals kept and then released away from the roost rarely found the carcass. Other individuals that were released at a new, previously unknown, carcass sometimes led roost-mates to it. As with the vultures, the recruiters gain by bringing other birds to a carcass so more can dismember it. The ravens have an additional incentive to arrive en masse: if there is a resident pair at the carcass, it can displace a few others, but the pair is overwhelmed if several from the roost are there together (Marzluff et al. 1996).

While the ravens are not known to make any deliberate recruiting signs, their roost site typically moves closer to the current carcass a day after it is discovered, by which time more birds know of it. The move from the prior roost takes place after dark; it is preceded in the last hour of light by ravens giving "social soaring displays" in which they fly high and may cover an area of 200 sq km. Ravens are attracted from at least 10 km to those that are soaring. This signals the coming location shift, which newly recruited individuals may associate with a new food source

(Marzluff et al. 1996). Even after the roost relocation, some that fed at the nearby carcass may go to other roosts and return the next morning to the carcass, followed by others from that roost (Heinrich 1994).

Studies in North Wales on an island with a stable roost of more than 1,500 subadult ravens in an extensive forest as well as smaller roosts of mated pairs found that ravens at the large roost were led to a carcass by the individual that had originally discovered it. The discoverers spent the night in the center of a part of the roost nearest the carcass; they were dominant at the carcass and were among the few from the roost that participated in the social soaring displays. In addition to being a signal that these birds know a place to feed tomorrow, the flights may also be a way for subadults to "show off" their food-finding ability, making them more attractive as future mates. At carcasses that had also been found by the local resident pair, the pair was able to defend it until the subadult discoverer recruited about five other birds. With each passing day until the carcass was depleted, additional birds—about six for every bird the previous day—came from the major roost. While the greater number may have reduced the feeding opportunities for the birds that were repeat visitors, in locales where snow may cover a carcass or where other scavengers such as canids may take a share, the diminishing balance of food would have been lost to them entirely (Wright et al. 2003).

Numbers at Roosts

The numbers of birds using communal roosts can be astounding. In addition to the sheer drama of seeing such numbers, these can provide useful information on actual populations of birds otherwise difficult to census. When the average distance that birds travel to reach a particular roost can be determined, this also helps indicate the likely density of the birds over suitable habitat during the day—bearing in mind, however, that individuals feeding together may head to separate roosts at the end of the day. The ecology and place on a food chain of different species means that what is a large roost for some will be paltry for other species. Top predators will never be as numerous on the ground or within flight range of any roost as will vegetarian or low-level carnivorous birds. And with the steady global decline of so many species, perhaps few of today's roost numbers will come near those from earlier decades.

Nonpasserines

The aggregations of shorebirds at postbreeding staging areas on the coasts of Alaska and Canada are among the best indicators of their total North American populations, which can most easily be counted at roost sites. Except perhaps for the smallest species, however, these populations are much lower today than before the era of large-scale market and sport hunting for sandpipers and plovers in the late nineteenth and early twentieth centuries that substantially contributed to the probable extirpation of the Eskimo Curlew (*Numenius borealis*) and the decline of the American Golden-Plover (*Pluvialis dominica*). On the Yukon-Kuskokwim Delta, Alaska, where 70,000 to 100,000 Dunlins feed at Angyoyaravak Bay, the flocks roosting during diurnal high tide number up to 21,500 (Handel and Gill 1992). On the Atlantic coast, the Bay of Fundy is the equivalent concentration point to Alaska's estuaries for shorebirds before they make long overwater flights. At Mary's Point, New Brunswick, a single roost of Semipalmated Sandpipers (*Calidris pusilla*) contained 350,000 birds on August 10, 1977; the average peak numbers at three sites around the bay at that season was 70,000 per roost (Hicklin 1987).

For raptors—never as numerous as most other classes of birds—a roost that has dozens, hundreds, or low thousands of birds is an impressive tally. Owls are less gregarious than many diurnal raptors, and most species are sedentary and territorial through the year, so reports of twelve Barn Owls (*Tyto alba*) in the same tree or old building during autumn or winter and as many as fifty in a small grove of trees are impressive (Eckert 1974, p. 12). For hawks, roosting aggregations usually occur outside the breeding season. The record for harriers is an estimated three thousand at a regularly used roost in the Blackbuck National Park, Gujarat, northwest India, where the majority are Montagu's Harriers (*Circus pygargus*), with perhaps 15–25 percent Pallid Harriers (*C. macrourus*), and a few Eurasian Marsh Harriers (*C. aeruginosus*). This may represent 1 percent of the minimum global population of

Montagu's Harriers. In North America, where there is only a single harrier species, the largest known roost of Northern Harriers (*C. cyaneus*) totaled 1,053 birds (Clarke et al. 1998).

Several falcon species have ecologies similar to swallows, in that they feed primarily on flying insects. Whether territorial or loosely colonial during the breeding season, some falcons, like swallows, form large flocks that roost together before and while migrating. The most impressive are those of the Amur Falcon (*Falco amurensis*), which breeds over much of northeastern Asia. Only in 2012 was it confirmed that perhaps the entire population funnels into northeastern India near the Burmese border, where in late October and early November termites from underground nests emerge from the forest floor to mate and disperse— becoming the prey that fuels the falcons on their subsequent flight across the subcontinent and Indian Ocean to Africa. Nagaland, where the greatest concentrations have been found, is hilly, mostly roadless, and densely forested, making it impossible to see the entire population gathered there. Observers at one vantage point at the edge of a reservoir have described hundreds of thousands in densely packed roosts, with perhaps a million or more in total in a single valley. Other major roosts have since been found in the neighboring states of Assam and Manipur (Weidensaul 2021, pp. 316, 339).

Among parrots, the species that usually feed in flocks may also form very large roosts where food is abundant and concentrated. Perhaps because the birds are difficult to count, most mentions of high numbers at roosts simply say "thousands," as for Alexandrine (*Psittacula eupatria*) and Rose-ringed Parakeets in India; "a minimum of 1,500" Yellow-lored Amazons (*Amazona xantholora*) in Mexico; and "many thousands" of Orange-winged Amazons (*A. amazonica*) in Guyana (Forshaw 1989, pp. 356, 360, 595, 628). In the state of Rio Grande do Sul, southernmost Brazil, roosts of Red-spectacled Amazons (*A. pretrei*) fluctuate in size, depending on the local food supply and available roosts within ten or more kilometers; one traditional roost has held as many as ten thousand birds (Belton 1984, p. 537).

For other large tropical forest birds, such as hornbills, that live in family groups that disperse widely each day to feed, a significant roost number is lower than for most parrots. Brown Hornbills (*Ptilolaemus tickellii*) move about in flocks of twenty to thirty (averaging 3.5 families) and join about as many others at a roost. In the Khao Yai National Park in Thailand, Wreathed Hornbills (*Aceros undulatus*) roosts may number one thousand or more (Tsuji 1996, pp. 66–67).

Perhaps no roosts have ever equaled in number and extent those of the now extinct Passenger Pigeon (*Ectopistes migratorius*) of North America. They preferred wooded swamps, which perhaps reduced access for some ground predators, and used other forests if wet ones were not available. Some roosts were traditional, used for months or years, then abandoned and recolonized a few years later. Some were said to cover 8–10 hectares. In 1812, Alexander Wilson estimated that the birds dispersed to feed every day as far as 96–128 km, and in 1831 Audubon noted that while the first birds might return to the roost at 4 p.m., the last did not arrive until midnight. Audubon wrote, "The noise which they made, though yet distant, reminded me of a hard gale at sea, passing through the rigging of a close-reefed vessel. As the birds arrived and passed over me, I felt a current of air that surprised me. . . . The Pigeons, arriving by thousands, alighted everywhere, one above another, until solid masses as large as hogsheads were formed on the branches all around. Here and there the perches gave way under the weight with a crash. . . . It was a scene of uproar and confusion" (Schorger 1955, pp. 78–79). Still earlier, the Reverend Cotton Mather wrote from Massachusetts to the Royal Society in London, probably in 1712, "Yea, they satt upon one another like Bees, till a Limb of a Tree would seem almost as big as a House. 'Tis incredible to tell, how Large & Strong & Many Limbs were broken down, by this New Burden upon ym [them]. The breaking of ym were heard at a mighty Distance. The Birds filled more than Half a mile, about from the Center, and the Noise they made, was like ye Roaring of the Sea" (Schorger 1938, p. 79). None of these observers tried to estimate the number of pigeons in the roosts, but Wilson and Audubon, elsewhere counting and timing migrating flocks that would have formed a single roost, respectively calculated around 2.2 billion and 1.1 billion birds. Other estimates later in the nineteenth century from farther west in the Passenger Pigeon's range exceeded 3 billion (Schorger 1955, p. 201–202). Of living pigeons, the Eared Dove

(*Zenaida auriculata*) comes next in numbers at a roost. In Argentina they roost in spiny thickets, dispersing during the day as far as 60 km to feed. One site occupying 350 hectares was estimated to contain "millions" (Skutch 1991, p. 27).

Passerines

Among swallows, eclipsing the 700,000 Purple Martins gathering before migration on an island in a lake in South Carolina as well as the 250,000 Southern Martins (*Progne modesta*) that for a few years came to the Plaza de Armas in Iquitos, Peru, are accounts of pre-migratory flocks of Bank Swallows (*Riparia riparia*) in England and of winter roosts of Barn Swallows in Africa. A roost of reedbeds in the Fenland area on the Norfolk-Cambridgeshire border held two million Bank Swallows in August 1968; most of these were juveniles, which wander after they fledge, while adults stick closer to their colony until they migrate (Mead and Harrison 1979). The Barn Swallows joining the multispecies roost in the Transvaal reedbed were estimated at 1–1.2 million (Rudebeck 1955). In Nigeria, Barn Swallow roosts have been found containing 1.5 million (Bijlsma and van den Brink 2005)

For swallows and for other birds that use roosts where conditions vary from year to year, numbers may more likely reflect the shifting conditions and availability of the roost sites than actual populations. In Botswana, rainfall and riverine water levels change dramatically from year to year. A roost in reedbeds on the Boteti River, which is fed by the Okavango Delta, held 2,000–100,000 Barn Swallows in 1992–93— when the water was high and many roosts in the region were used—and two to three million the following year when fewer roosting sites were available because water was low. In dry years fewer insects are produced, so swallows have to travel farther from the remaining roosts to find them, leaving the birds in poorer condition, with fewer surviving the season. These high numbers may therefore indicate lower overall populations (van den Brink et al. 2000).

Many corvids that feed on the ground in open country gather in large winter roosts that likely combine resident and migrant birds. These are the agglomeration of roosts that may have begun as early as the previous July, when local, postbreeding birds begin to assemble; they combine into large roosts during autumn and into still larger ones in winter. In the Ythan Valley in Aberdeenshire, Scotland, a winter roost was estimated to contain sixty-five thousand Rooks. In Uppsala, Sweden, a winter roost of Jackdaws held forty thousand birds (Coombs 1976, pp. 102, 125). At Drum Island, on the Cooper River in Charleston, South Carolina, some forty-five thousand to sixty thousand Fish Crows (*Corvus ossifragus*) have roosted together in November, with the numbers dropping to eleven thousand in midwinter (McNair 1988).

Thrushes in the genus *Turdus* are territorial when they breed, but they assemble in communal roosts soon afterward; newly fledged young are the first birds, led by failed breeders, followed by males that may or may not also be tending nests in nearby territories, and finally the adult females and their final brood. In Cambridge, Massachusetts, William Brewster (1890) kept track of several roosts of American Robins in the area. On the evening of August 4, 1875, he estimated an exceptional twenty-five thousand birds coming to a wet woodland at Little River, Arlington, noting, "I feel sure this was far below the actual number." The numbers at these summer roosts, large and small, tapered off in September and the last of the robins were usually gone by early October, a far different schedule from that of today. In Arkansas in the 1920s, robins first arrived at a roost in the Ozarks in late October, with the numbers swelling into December, when, on December 16, 1928, there were at least 250,000 roosting in 2.6 square km of woodland; the birds departed after they had depleted the fruit on trees within several kilometers (Black 1932). In Britain, some winter roosts of Fieldfares (*T. pilaris*) and Redwings (*T. iliacus*) number more than 20,000 birds combined, and in January 1975, a roost of 200,000 Fieldfares with some Eurasian Blackbirds (*T. merula*) was found in northern France (Clement and Hathway 2000, p. 48).

A survey of wintering Dickcissels (*Spiza americana*) in the rice- and sorghum-growing parts of central Venezuela found concentrations in different areas of one to six million. In April, when birds were beginning to depart for North America, sixteen roosts in sugar cane fields ranged in size from 20,000 to 2,950,000 (median 580,000), with nearly 40 percent of all roosts larger than one million birds. Some roosts, therefore, can contain as much as 30 percent of the known Dickcissel population (Basili and Temple 1999). For the Dickcissels the feeding areas and most convenient roost sites are likely to be the same winter after winter, if the same crops are planted. In Europe wintering Bramblings, which feed on beech mast, search over long distances to find a good crop, and their roosts inevitably shift to be nearby. After feeding in the leafless beech trees, Bramblings commute to much more sheltered conifer groves. In years of extensive mast crops, some roosts in Switzerland have held numbers estimated at 11 million, 50 million, and 70 million birds (Newton 1973, pp. 28–30). Among the Old World sparrows, postbreeding roosts of House Sparrows (*Passer domesticus*) of 19,000 have been recorded in London and 100,000 near Cairo; Golden Sparrows (*P. luteus*) in Senegal have roosts of 400,000 and one million; and mixed House and Willow Sparrow (*P. hispaniolensis*) roosts in Rajasthan, India, number over one million birds (Summers-Smith 1988, pp. 140, 52, 172).

The winter roosts of some New World blackbirds surpass most of these numbers. Great-tailed Grackle (*Quiscalus mexicanus*) roosts in sugar cane fields in the Lower Rio Grande Valley of Texas may number 500,000 birds (Johnson and Peer 2001). In Evangeline Parish, Louisiana, the forests covering 1,750 hectares in a lake impoundment attract millions of blackbirds and starlings. Based on 1980s Christmas Bird Count data, the numbers of each species fluctuate yearly, with high counts in late December including 10 million starlings, 27.5 million Common Grackles, 53.1 million Red-winged Blackbirds, and 38.2 million Brownheaded Cowbirds. The highest total number of birds recorded there is 108.7 million. Most of the birds leave the roost by late February, decreasing from a collective 18 million in mid-February to 6,300 Redwinged Blackbirds in mid-April (Brugger et al. 1992).

Summary

In contrast to the majority of birds, which roost singly, in family groups, or small aggregations, some species roost seasonally or year-round in large flocks. Communal roosting among large numbers of unrelated birds is most widespread after the breeding season, when many birds, even essentially sedentary ones, are no longer strictly territorial. Migrants such as shorebirds, swallows, and other birds that move about in flocks from the time they leave their breeding territory until they return the following year often roost in flocks. Some other migrants, whether en route or at their wintering destination, spend the day foraging solitarily, even maintaining a daytime territory, but they also gather in flocks for the night. Across the spectrum of annual and life cycles of various species, the individual birds most likely to roost in flocks are sexually immature, unpaired, not holding a territory—at least one that requires constant presence—or migrating and wintering at a distance from their breeding grounds. The major benefits of roosting in large aggregations include lower thermoregulatory costs and reduced predation risk.

Large flock roosts may consist of a single species, or several species closely related taxonomically and overlapping ecologically, or of species whose diurnal lives are entirely different. Each of the various species may be there year-round or only seasonally in greater or lesser numbers. Use of the roost may vary by age class, breeding status, and sex of each species. At large communal roosts, there is often a hierarchy, within or among species, based on age, gender, and condition, with dominant birds securing and retaining the more favored places. Some places are superior because they are closer to feeding areas or provide more thermal benefits, ease of landing and takeoff, greater visibility for spotting predators, or avoidance of droppings coming from neighboring birds.

Communal roosts may also increase foraging efficiency by bringing together more birds that can search successfully for food the next day, especially if food is patchily located, whereas solitary birds might have less chance of finding it. In some species, individuals returning to a roost after a day of poor foraging may the next morning watch other birds depart that appear to be going directly to a place where they fed the day before. While this information is most useful to conspecifics, at mixed roosts occupied by ecologically similar species, these birds may also be alert to the behavior of all their neighbors. In most cases where roosts may serve as information centers, there is no particular benefit received by the knowledgeable birds departing in the morning for a place where they expect to continue finding food. But among scavengers such as vultures and ravens, recruiting more birds to a newly discovered carcass facilitates tearing it apart as well as providing vigilance against predators and competitors.

The numbers of birds using communal roosts can provide useful information on actual populations of birds otherwise difficult to census. Some roosts may harbor a significant share of a species' local or global population. The ecology and place on a food chain of different species means that what is a large roost for some will be paltry for other species. Top predators will never be as numerous on the ground or within flight range of any roost as will vegetarian or low-level carnivorous birds. And with the steady global decline of so many species, perhaps few of today's roost numbers will come near those from earlier decades. At one extreme, nineteenth-century Passenger Pigeon roosts were sometimes estimated to contain three billion or more birds, while today the largest known roosts, of several blackbird species and European Starlings, have totaled 108.7 million. Winter Brambling roosts may hold as many as 70 million birds, and Amur Falcons in northeast India may on migration roost in flocks of millions. Roosts of winter Dickcissels in Venezuela may include 1–6 million birds, and some swallow roosts hold 1.5–2 million. At the other end of the spectrum, fifty Barn Owls together or a thousand Northern Harriers is an impressive gathering.

7

Roosting Times

Several factors influence the time that birds approach and depart their roost. For diurnal birds, light level is probably the most important factor, because this affects their abilities to feed and to find their way to their roost, which may be at some distance from the last place they foraged. In addition to being able to see their route to the roost—as well as from the roost to their first morning destination—many birds, especially those moving alone, are also concerned about how visible they may be to predators. Some crepuscular and nocturnal birds also time their travel to and from roosts to avoid predators, while for other nocturnal birds their schedule is set by the activity periods of their prey. Lunar cycles and weather conditions, as well as extremes in temperature, further influence the time birds spend roosting.

Effects of Light Level and Day Length

Light level and day length are closely related, but each may independently affect the time spent roosting. Where spring and summer days are long, many birds return to their roost at higher light levels, even when they have young to feed, than during fall and winter, when they must use all available daylight. In the darker seasons, many birds typically leave their roost in the morning at lower light levels (than during spring and summer) to get an earlier start on foraging during the short days. Shorebirds dependent on tidal cycles, and even other

birds normally thought of as diurnal, sometimes use hours of darkness as well. Chronotype, the timing of an animal's activities over the twenty-four-hour cycle, thus varies throughout the annual cycle as well as among individuals within any species, reflecting age, sex, fitness, latitude, and probably other intrinsic and extrinsic features that can only be pinpointed in long-term studies with individually known birds.

The time and the manner birds arrive at and depart from their roost are best known for those that roost singly but consistently in the same place, such as cavities, where they can easily be monitored, and for species where this is conspicuous, such as birds that roost in flocks at well-established locales. Other territorial birds that can be followed relatively easily also provide some information, while for the many birds with sleeping and waking habits harder to observe, the times when they first vocalize or are first visible feeding may serve as proxies for when they leave their roost site in the morning.

Species Roosting in Cavities

Most cavity roosters are territorial, so they are relatively close to their roost all through the day. They typically enter their roost earlier than many other birds and remain in them longer. Since cavities are scarce and subject to competition both within and among species, it is likely that individuals seek to occupy them early even if this requires cutting short the feeding day. At a time when many nearby birds that roost in foliage are still active and singing, Ochre-collared Piculets (*Picumnus temminckii*) in the Atlantic forest of Brazil enter their roost holes an average of thirty-nine minutes before sunset. The piculets sometimes look out of their cavity or vocalize within it; their appearance at the entrance must indicate to any other birds seeking a cavity that this one is occupied. In the morning, the first piculets emerge about eighteen minutes after sunrise (Bodrati et al. 2015).

Other woodpeckers are also known to be leisurely in the morning. Guadeloupe Woodpeckers (*Melanerpes herminieri*), endemic to that Caribbean island where no birds need wait for temperatures to rise, typically first look out of their hole fifteen minutes after sunrise, spending another two to eleven minutes there before emerging (Villard 1999, p. 61). The last closely observed Ivory-billed Woodpeckers (*Campephilus principalis*), in the Singer Tract in Louisiana, would emerge from their roost holes about fifteen minutes after good daylight, when the sun had already lighted the treetops. In March, they arose around 6:30 a.m., in May and June from 4:45 to 5:15 a.m., and in December from 6:45 to 7:15

Woodpeckers large and small typically emerge from their roost hole later in the morning than most other birds begin their day.

a.m. Each bird would emerge silently from its cavity, climb to the top of the tree, and sit, stretch, and preen before calling to locate its mate and then join it (Tanner 1942, p. 57).

Weather also affects woodpecker roosting time. On clear days in East Kalimantan, Indonesia, twenty to thirty-eight minutes before sunset, family groups of Grey-and-buff Woodpeckers (*Hemicircus concretus*) arrive at the trees where they have excavated many cavities. They may immediately enter cavities or spend up to thirty-one minutes inspecting various ones, entering some, and occasionally fighting over them. Once settled in a cavity, the birds often spend several minutes looking out. On dry days, the last individual in one group of four, an adult male, entered a cavity approximately sixty minutes before sunset. On days with heavy rain, however, the birds arrived earlier and all birds were in their roosts forty to sixty minutes before sunset. On a dark afternoon with a thunderstorm, the group returned 112 minutes before sunset, sat out the storm in cavities, and, thirty-five minutes later, spent an hour chasing around the cavities and further excavating them (Lammertink 2011).

The toucans and barbets that roost in cavities likewise typically do not emerge until it is fully light, and they remain longer, sometimes another hour, on cloudy or rainy mornings. Fledglings still with their parents tend to remain in the roost longer in the morning and to return to it earlier in the evening than adults, which also enter their cavity well before dark. The barbets that live in larger social groups, and often roost together if larger cavities are available, begin duetting and chorusing in the morning as soon as the last of them has left the cavity (Short and Horne 2001, pp. 60, 195).

Where nest or roost sites are scarce, birds may adjust their entry time to the specific locale or features of the site. In Sinaloa, Mexico, two nearby nests of Orange-fronted Parakeets (*Eupsittula canicularis*) in arboreal termite nests show how placement and light levels affect the time available for other activities. At a nest in a ravine and with an entrance hole facing downward, the birds returned to it shortly after 5 p.m., perched near it, and entered it at 5:30, about when the sun disappeared from their view. At another nest higher on a hill, the birds did not enter until 6:18 to 6:30, when, from that vantage point, the sun went below the horizon (Hardy 1963).

In one of the most comprehensively studied of all cavity-nesting and roosting birds, the Great Tit (*Parus major*), waking and roosting times vary over the course of the year, depending on what the birds of each sex are doing. During the nesting season in the Netherlands, the male emerges before sunrise, as many as forty-five minutes earlier than the female; he goes to the nest cavity and sings from outside it, and the female vocalizes from within. When females are laying eggs and incubating them, they leave the nest shortly after sunrise or sometimes up to thirty minutes after. When the young have hatched, females leave several minutes before sunrise. At the end of the day, females return to their nest cavity on average sixty-four minutes before sunset during the egg-laying phase and one hundred minutes before when actually incubating. When feeding that first brood they do not retire until after sunset. When laying a second clutch, near the summer solstice, the female returns to the nest 2.5 hours before sunset; when this clutch hatches, she finishes feeding the brood before sunset. Meanwhile, males have shifted to sleeping in the open, so their precise roosting time is less well known, but it is a few minutes later than the females'. In autumn and winter, the waking and roosting times of the sexes are more aligned, with the male still having a slightly longer day. Then, even in December, however, Dutch males begin their day at higher light levels than in the breeding season, an average of twenty-six minutes before sunrise—although later on rainy mornings—while at 67°N in Finland, Great Tits emerge nearly two hours before sunrise. During December in the Netherlands, both sexes go to roost about five minutes after sunset. Later in the winter, they roost about ten minutes before sunset, even as the winter days grow longer; this is much later relative to sunset than during the breeding season (Kluijver 1950).

Slightly farther south, in Germany, where the spring and summer days are a bit shorter, a population of Great Tits was monitored for the time females first left their nest during April, when they were incubating; this ranged from eight minutes before sunrise to forty-three minutes after sunrise. The birds that laid their first egg earlier in the season consistently left their nest earlier in the morning. Other studies found that individuals were consistent year after year in their emergence time and that males that waken earlier and begin singing earlier in the morning mate with females that then begin nesting earlier in the season. Still other research has shown that older individuals tend to initiate nesting earlier in the season than young birds, but whether this means they wake up earlier in the morning has not been demonstrated. Nor are there data on whether, within a population, earliest risers during the incubation stage continue the habit at other times of year. But among Dark-eyed Juncos (*Junco hyemalis*), which nest in sheltered open locations, not cavities, the females that rise earliest in the morning are also the earliest in the season to begin laying eggs. At this season female juncos leave their nest over a wider span of light levels than the Great Tits, from forty-one minutes before sunrise to 117 after sunrise (Graham et al. 2017).

A study of Blue Tits (*Cyanistes caeruleus*) found that the time leaving the nest cavity in the morning varies significantly, even within a brief period in the breeding season. Until about five days before egg laying began, females emerged an average of 26.7 minutes after sunrise. In the last two days before egg laying, however, they left the nest prior to sunrise, as many as 5.5 minutes before it. While similar early morning behavior in Great Tits has been shown to result in more extra-pair copulations, that was not found in this population of Blue Tits (Schlicht et al. 2014).

At 67°N in Finnish Lapland, female Siberian Tits (*Poecile cinctus*) enter their nest 220 minutes before sunset when they have not yet laid any eggs; 270 minutes before when laying; 220 minutes before when incubating; and eighty minutes before when feeding nestlings. During this period, the days are rapidly growing longer; the birds are therefore spending several more hours out of the nest. Their rising time is more consistent, fifty to seventy minutes after sunrise. At this season males leave their roost earlier, sometimes waiting outside the female's cavity thirty minutes for her to emerge, and retire slightly later. In January at the same locale, the tits' total time out of the roost was about five hours (Hailman and Haftorn 1995). In May, Willow Tits (*P. montanus*) at the same latitude begin their day at 1 a.m., while at 48°N they rise at 4:15 a.m., but at both latitudes they retire around 8 p.m., long before sunset (Armstrong 1954).

Species Roosting in Exposed Sites

Birds that do not roost in cavities also leave their roost in the morning at lower light levels than they return to it at the end of the day. This is true both for birds at higher latitudes during autumn and winter—when nights are longer and colder than the days, and birds are eager when they wake to replenish their energy reserves—and for birds where the length of the night is the same all year. In England, Pied Wagtails (*Motacilla alba*) use a communal roost through most of the year. In July, when they have finished nesting, they arrive by twenty minutes after sunset and some are still asleep the next morning fifteen minutes after sunrise; in winter, the wagtails all arrive by thirty minutes after sunset, which is hours earlier than in July, and depart by sunrise, then much later than in July (Broom et al. 1976). Similarly, in Singapore, just above the equator, roosting flocks of Common Mynas (*Acridotheres tristis*) depart in the morning at lower light levels than they return in the afternoon, and the roosts break up much more rapidly than they assemble (Nee and Yeo 1993).

For some birds whose roost sites are not easily found or monitored, first singing can be a proxy for time waking up or leaving the roost. It may also reveal the sequence in which various species in the same habitat begin their day. In both the tropics and higher latitudes, many birds start singing when the sky is still entirely dark, while others commence after sunrise but before there is enough light for foraging. It seems likely that these birds begin soon after they wake up, at least before they undertake any other activity. The same applies to crepuscular birds like nightjars that begin singing when light decreases to a certain level. The actual time, of course, that each bird initiates song shifts daily if song is aligned with light level. The same species at different latitudes will therefore likely first sing at different times on the clock, but probably at similar light levels. In addition, different individuals may have their own schedule, based on light levels at their roost, as well as varying by age and condition.

A study running from January 15 to September 1 in Washington, D.C., used civil twilight, when the sun is 6° below the horizon, to track the light level and sequence of species first singing. The Eastern Kingbird (*Tyrannus tyrannus*) consistently began singing fifteen to twenty minutes before civil twilight began, which is forty-two to fifty minutes before sunrise; the Wood Thrush (*Hylocichla mustelina*) at the beginning of civil twilight; the House Wren (*Troglodytes aedon*) twelve to fifteen minutes later. Still earlier, the American Robin (*Turdus migratorius*), Northern Cardinal (*Cardinalis cardinalis*), and Song Sparrow (*Melospiza melodia*) first sang

within the period of astronomical twilight, when the sun is between 18°
and 6° below the horizon. Cloudy mornings retarded the initiation of
song for each species, but temperature did not have an effect, even as the
seasons changed from winter to summer (Allard 1930).

A comparable study at Barro Colorado Island, Panama, tracking
the initiation of song by twenty-seven species of fifteen families in nine
orders, found that singing peaked at 12.5 minutes after sunrise, with
Collared Forest-Falcons (*Micrastur semitorquatus*) being the first, beginning twenty-five minutes before dawn (unfortunately not defined or
quantified in terms of light level), and Red-capped Manakins (*Cerato-
pipra mentalis*) the last, at fifteen minutes after dawn. Looking for ecological correlations that might explain the sequence, the researchers
found that species with songs in the same frequency range as nocturnal
insects did not begin singing until those insects had stopped. Birds singing in the same range as other birds did not inhibit any species (Stanley
et al. 2016). Whether the sequence of singers reflects their actual waking
time is not known, but understory birds like the manakins would have
difficulty foraging before the sun rises high enough to bring light to the
forest interior, while upper story and canopy dwellers could begin foraging earlier.

The sequence of initiation of song by fifty-seven passerines in Britain,
Switzerland, and Portugal was found best correlated with their eye size
relative to their body mass. Species with relatively larger eyes began
singing earlier in the morning; smaller birds began to sing at lower light
levels than larger birds of the same eye size. Smaller birds might generally be expected to become active earlier in the morning because they
have depleted a larger share of their energy reserves during the night.
The eye size correlation, however, suggests that birds with different foraging methods that require different levels of visual acuity may be able
to start their activity at different light levels, possibly with a bout of song
before they begin foraging (Thomas et al. 2002).

The urge to leave the roost at lowest light levels possible must be bal-
anced with the risk of predation, when both nocturnal and diurnal predators may be active. In British Columbia, wintering Dark-eyed Juncos with
sixteen hours of darkness arrived earlier in the morning at feeding sites

sheltered from predators than at exposed sites. Individuals experimentally held the previous day for several hours without food were more willing to feed the next morning at exposed sites where they were at greater risk. While all birds arrived later on overcast days, the actual light levels were nevertheless lower than at their arrival time on clear days, indicating a compromise between predator avoidance and hunger; on the cloudy days, the juncos were also more vigilant when feeding (Lima 1988). Similarly, at a communal roost of Greenfinches (*Carduelis chloris*) in Denmark, birds left the roost earlier when more were present, perhaps because this increased their safety; they left later when light levels were low, and also later when Sparrowhawks (*Accipiter nisus*) were nearby, since these often hunt at low light levels (Kiis 1986).

Impact of Distance from Roost to Feeding Sites

Birds that disperse widely from their roost to feed during the day usually depart early in the morning, traveling as soon as they can see well enough. Their return time, however, reflects their feeding success and distance traveled. Those that have fed well may return to the roost at higher levels of daylight than birds that have had less success and continued feeding as late as they can. Among wintering flocks of Common Cranes (*Grus grus*) in Spain, birds return later to their roost when food is scarce, but during the phases of the lunar cycle when the moon rises before dark, they extend their feeding time, by as much as an hour when the moon is full (Alonso et al. 1985).

In Scotland, Northern Harriers (*Circus cyaneus*) roost in small groups on boggy land, dispersing in every direction each morning; most return in a period of forty to sixty minutes, starting fifteen to thirty minutes before sunset and continuing, with the very last birds, until forty-five minutes after sunset. On wet and stormy days, most arrive earlier than on fine evenings, but if the weather clears at the end of a rainy day, birds come in later, taking advantage of the clearance for late hunting. Most harriers depart before sunrise (Watson 1977, p. 243–244). In a deforested area of southern Guatemala, Crested Caracaras (*Polyborus plancus*) come from cattle pastures to roost in one of the few remaining large trees;

when the sun sets at 6 p.m., they arrive between 5:20 and 6:45, with the bulk coming between 6:15 and 6:35 (Johnson and Gilardi 1996).

Among parrots, arrival and departure times range widely among the species in which this has been observed. This may reflect the ecology of each species. Those that live in forests and forage and roost below the canopy find themselves in darkness earlier than canopy feeders and species of open habitats. Parrots that disperse far from their roost each day may also return later. Thick-billed Parrots (*Rhynchopsitta pachyrhyncha*), which live in highland pine forests in Mexico, roost—wherever they are still common—in large flocks. They begin returning to their roost three hours before sunset, with the bulk arriving in the final hour; they also return earlier ahead of afternoon thunderstorms. In the morning, the entire flock departs at first light. African Grey Parrots (*Psittacus erithacus*), in contrast, begin their return to a communal roost toward dusk, with some birds not arriving until well after nightfall; they leave at sunrise (Forshaw 1989, pp. 464, 314).

Swifts disperse great distances from their roost to feed; they show a wide variety of roosting times. In Costa Rica, White-collared Swifts (*Streptoprogne zonaris*), which nest and roost behind waterfalls, begin appearing in the general area around midafternoon and begin entering their roost at 4:30 p.m., earlier on dark days (Marin and Stiles 1992). Edible-nest Swiftlets (*Aerodramus fuciphagus*) nest and roost in caves and have evolved echolocation to navigate these in the dark. Echolocation may also be deployed when feeding, because they feed intensely late in the evening. In the Andaman Islands, the peak roosting hours are 5 to 8 p.m., with nearly 93 percent returning after dark. More return later on moonlit nights, when they may be foraging longer or farther away. In the morning, they leave their cave before dawn. In this way, they avoid the two major hunting times of local owls, to which they would be most vulnerable when concentrated at the cave entrance (Mane and Manchi 2017). This early departure contrasts with that of most swifts, which, like many other birds roosting in sheltered locations, emerge later than species sleeping in the open. Some swifts, especially in cooler latitudes and on dark or wet days, do not become active until even later than cavity roosters, because fewer insects are flying in those conditions. Common Swifts (*Apus apus*) nesting at high latitudes emerge later relative to daylight than those at lower latitudes; this may still, however, give them more hours of daylight available for foraging. Those with young to feed leave their roost later than those without, presumably waiting until aerial insect densities have increased sufficiently to justify the effort (Chantler 2000, pp. 14, 30).

At mixed-species roosts, the timetable of each species may be different. During one August and September at a heronry in Cape May County, New Jersey, where the roost was far from feeding areas, the sole Tricolored Heron (*Egretta tricolor*) arrived 52–34 minutes before sunset; Little Blue Herons (*E. caerulea*) 17–15 minutes before; Great Egrets (*Ardea alba*) 15–10 minutes before; and Green Herons (*Butorides virescens*) 7–14 after sunset. The bulk of the Black-crowned Night-Herons (*Nycticorax nycticorax*) left the roost 18–23 minutes after sunset. In the mornings, the night-herons returned 30 minutes before sunrise, the Green Herons began leaving soon after, and most of the remaining birds of all species left 6–12 minutes before sunrise. The entire departure took only 20–25 minutes while the arrivals were spread out over more than an hour (Seibert 1951). Similarly, at blackbird roosts arrival times differ by species, and also by age and sex. At a late summer roost in Maryland, subadult male Red-winged Blackbirds (*Agelaius phoeniceus*) were consistently the first to arrive, followed by females and juveniles. Adult males were sporadic in their time of arrival and were followed by Bobolinks (*Dolichonyx oryzivorus*), European Starlings (*Sturnus vulgaris*), Common Grackles (*Quiscalus quiscula*), and finally Brown-headed Cowbirds (*Molothrus ater*) (Meanley 1965).

In contrast, birds that have a reliable and abundant artificial source of food available close by and at all hours may develop a consistent pattern of arrival time at their roost, since they do not have a long commute, any need to search distantly, or a fluctuating supply. The factors influencing their arrival time may therefore be easier to identify. Herring Gulls (*Larus argentatus*) that feed at dumps in Maine are satiated earlier during summer months, when they can begin feeding earlier; they return to their roost well ahead of sunset. In autumn, however, the flight does not begin until it is nearly dark. In summer, there is no difference in arrival time between clear and cloudy days, but in autumn more birds arrive earlier on cloudy days (Schreiber 1967). A roost of Common Ravens (*Corvus corax*) in Thuringia, Germany, that in winter held as many as 574, exploited a nearby composting facility. The ravens similarly returned to the roost later on cloudless evenings and, relative to sunset, on days with fewer hours of daylight. Most birds arrived within a thirty-minute period that spanned twenty minutes before sunset and ten minutes after. Weather conditions, including temperature, wind speed, and precipitation, did not affect feeding success or roosting time, and the birds did not continue feeding longer on moonlit nights (Janicke and Chakarov 2007).

Effects of the Moon

A large or full moon rising when daylight fades maintains light levels high enough for some birds, like Common Cranes, to continue feeding and delay flying to their roost (Alonso et al. 1985). If moonlight begins later at night, after birds have gone to roost, some may resume activity. Nocturnal birds that hunt most often at dusk and before dawn may set forth again. Some diurnal birds simply sing from their perch; others may leave the shelter or concealment of their roost for other activities.

Barnacle Geese (*Branta leucopsis*) wintering in the Netherlands sleep about two hours less during long nights with a full moon; in summer, when sleep is spread across more of the entire twenty-four-hour cycle, the moon has less impact, but in both seasons there is less sleep during the day when the moon is full and visible at night. During these brighter nights, geese may spend more time feeding (van Hasselt et al. 2021). Similarly, in European Starlings, nighttime sleep waxes and wanes with the moon, about two hours less when the moon is full than when it is new, as shown with birds in cages (van Hasselt et al. 2020), but it is likely that in natural situations the starlings nevertheless remain at their roost through the night even while awake.

Male Houbara Bustards (*Chlamydotis undulata*) on Lanzarote island in the Canaries display on nights with a full moon, giving their booming calls twice as frequently as by day. In addition to providing better sound transmission and less acoustic competition, both of which are available on any still night, the nights with a full moon can give females better visibility conditions to assess the males, while any copulations that may follow are less likely to be interrupted by rival males than during the day. The bustards on Lanzarote have no nocturnal predators, so enhanced predator detection is not a factor there (Alonso et al. 2021).

The extent to which the moon affects roosting habits and timetables may depend not only on its fullness but also its distance from Earth. In every lunar cycle, the moon moves in its orbit from apogee—its most distant point from the planet—to perigee, approximately two weeks later, bringing it as many as 46,000 km closer and making it appear significantly larger in the sky—a "supermoon." At perigee, the moon is 30 percent brighter. A physiological study of Barnacle Geese in Scotland when perigee coincided with a full moon for three months in succession found that the birds raised their body temperature and, to a lesser extent, heart rate from the levels they maintained during sleep at other parts of the lunar cycle. The geese roosted on water and did not leave to forage, but their likely greater alertness may be valuable when more nocturnal predators are active on the brightest nights (Portugal et al. 2019).

Far less frequently and with no regular pattern, the moon is aligned between the Earth and the sun, creating a solar eclipse. This disrupts the day for most birds, sending them back to roost, even when true darkness never comes. During a partial solar eclipse in Dutchess County, New York, on August 21, 2017, that occurred from 1:23 to 2:44 p.m. and reduced light levels 50–75 percent, Turkey Vultures (*Cathartes aura*) began returning to their roost a few minutes after the eclipse began. More came while the eclipse was well underway, and a few not until it was already over. Some of the vultures left the roost between 3:38 and 3:58 p.m., but others remained and still more arrived within the next hour, well before their normal August roosting time (Platt and Rainwater 2018). In an annular eclipse on May 10, 1994, when there was a ring of sun around the moon and light level decreased by 80 percent, herons of four diurnal species at a colony near Wichita, Kansas, returned to roost soon after the eclipse began at 10:30 a.m., peaking when the sky was darkest, while none of the colony's Black-crowned Night-Herons departed (Maccarone 1997).

During a total eclipse on February 26, 1998, at a bay on the coast of Venezuela, Royal Terns (*Sterna maxima*) stopped feeding and disappeared thirty-nine minutes before totality. At thirteen minutes before totality, when the light conditions were like those just before sunset, Magnificent Frigatebirds (*Fregata magnificens*) and Brown Pelicans (*Pelecanus occidentalis*) also left the bay, the frigatebirds flying inland while the pelicans went to roost on cliffs bordering the bay. When the eclipse was total, only Laughing Gulls (*Larus atricilla*) were active, flying back and forth over the water in a tight flock. Twelve minutes after the solar disc began emerging, the first frigatebirds and pelicans returned to the bay; soon after, they and the gulls resumed feeding, while the terns had not yet returned ten minutes after the sun was completely visible (Tramer 2000).

Effects of Temperature and Extreme Weather

In addition to light levels, temperature can affect the time birds leave and return to their roost. In some situations, the two are linked directly, as when on cold, dark, and rainy mornings birds roosting in cavities wait to emerge until at least slightly later. Weather conditions associated with cold may also affect the availability or accessibility of food. Pygmy Nuthatches (*Sitta pygmaea*) sometimes remain in their tree cavities forty-two hours—two nights and one day—if snow covers the branches and needle clusters on which they forage (Kingery and Ghalambor 2001).

When cold does not affect their ability to find food, some birds may nevertheless during extreme cold spend more of the day at their roost even when it is bright—and some birds do not leave it all day. Canada Geese (*Branta canadensis*) wintering in Illinois, depart their roost later in the morning when the temperature is −6°C than they do on warmer days. When it is less than −9°C at sunrise, most do not leave their roost at all. They then spend most of the day sleeping, with their bill placed under the scapular feathers and their feet drawn into the flank feathers. In this temperature range, geese can survive fifteen to twenty days without feeding. So, to spend one or a few days at the roost, if the temperature does not rise later in the day, is the more energetically efficient course (Raveling et al. 1972). During winter, Ring-necked Pheasants (*Phasianus colchicus*) in Finland typically descend from their roost in trees for only twenty-five to ninety minutes per day. In extreme weather, the benefits of searching for food may not compensate for the energy expended; pheasants may instead remain on their perch for at least forty-two hours (Cramp 1980, p. 509).

Other birds do leave their roost in extreme cold, but they adjust their schedule. Merlins (*Falco columbarius*) wintering in Saskatchewan normally depart their roost in spruce trees before sunrise and return after sunset, but when the temperature is lower, they leave later and return earlier (Warkentin 1986). Black-billed Magpies (*Pica hudsonia*) near Edmonton, Alberta, usually leave their communal winter roost in the forty-five minutes before sunrise; their return time is less consistent, but, like many birds, later relative to sunset on the shortest days of the year. On cold mornings, when the temperature is less than −19°C, they depart slightly later, and on cold afternoons, more frequently return early (Reebs 1986). Mourning Doves (*Zenaida macroura*) in New Brunswick, Canada, at the northern end of their range, leave their roost slightly later on cold days and return more significantly later. With a body mass smaller than that of the magpie, the Mourning Doves lose heat more rapidly, so they need to feed as long as possible before the long Canadian night; this supersedes any benefits from expending less energy by roosting (Doucette and Reebs 1994). A nocturnal bird, the Black-crowned Night-Heron (*Nycticorax nycticorax*) also returns to its roost earlier in cold weather. In California, it usually comes back to the winter roost around fifty-six minutes before sunrise if the temperature is 7°C or higher, but several minutes sooner at lower levels (Perlmutter 1992).

Under extreme conditions, the same energy considerations may influence birds with even a reliable food source. Where Black Ducks (*Anas rubripes*) were fed daily at a piece of open water on the Rideau River near Ottawa, Ontario, on mornings when the winter air temperature was less than −25°C, the ducks left their roost 100 minutes later and returned 160 minutes earlier than on days when it was more than −2°C. Cloud cover did not affect the departure and return times. The temperature was the same at their feeding site—a flight distance of eleven minutes—but the wind speed at the roost site was 8 km/h less, reducing heat loss. In addition, at the roost the ducks huddled together and tucked their bills into their scapulars, further reducing heat loss in ways they could not have done while feeding (Brodsky and Weatherhead 1984).

Even some very small birds remain at the roost longer in harsh conditions. Near Fairbanks, Alaska, where Black-capped Chickadees (*Poecile atricapillus*) roost in the open, they are much less active when the temperature is −46°C to −51°C than when it is −28°C. On these cold, overcast days, they come to feeders later in the morning and retire earlier for the night than on warmer, clearer days. Some do not come at all. For them as for much larger birds, it is sometimes more efficient to remain at the roost all day than to expend energy searching for food or even traveling to a known source (Kessel 1976).

At lower latitudes and during any season, aerial insectivores similarly wait at their roost during colder weather for the air to warm up enough for many insects to take wing, which in some places is above 18°C with numbers peaking at 25°C. Chimney Swifts (*Chaetura pelagica*) nesting in Illinois begin calling in their roost thirty minutes before daylight. On many mornings when swifts depart, typically eleven minutes before sunrise, some reenter their chimney within thirty minutes of their initial departure, presumably having not found enough insects aloft to justify their effort. On cold or rainy mornings, 90 percent of the swifts reenter within thirty minutes, remaining until later in the morning or until the rain ends (Zammuto and Franks 1981). In Baroda, India, the time during December and January that House Swifts (*Apus affinis*) first call from within their nest and then leave has been correlated with temperature; the difference is as many as 139 minutes, with first calls before sunrise and departure at 10 a.m. or later, depending on the temperature. On cold mornings, swifts leave their nest later, but their first vocalizations are independent of sky conditions. Nor is it the absolute temperature that determines departure, but the rise in temperature. Whether the early morning temperature is 11°C or 20°C, House Swifts do not leave their nest until they detect that it is warming; on cooler days, the rise has to be greater than on warmer ones, as it must also be on cloudy mornings (Razack and Naik 1965).

Large soaring birds, for which mobility is limited to the hours that warm air creates thermals, must wait until air temperatures reach certain levels. In winter, Cape Griffons (*Gyps coprotheres*) in southern Africa often stay at their roost until 10 a.m., when sufficient thermals may have developed from the cold ground, and return four hours later. One marked bird in South Africa had an average winter departure time of 11:38 a.m. and usually was abroad for only 2.5 hours; in summer, the period extended by about ninety minutes (Mundy et al. 1992, p. 81). Turkey Vultures at the Malheur National Wildlife Refuge in Oregon depart their roost three to four hours after sunrise and return 2.5 hours before sunset. On breezy days, they leave earlier, but when there is rain or snow some do not leave at all (Davis 1979). Similarly, in Africa Marabou Storks (*Leptoptilos crumeniferus*) often wait at their roost for several hours after sunrise for thermals to develop (Hancock et al. 1992, p. 136).

Elsewhere, for reasons that are not evident, some birds may remain at their roost long after weather conditions seem favorable to begin the day's activities. At Raine Island, Australia, 11°S and east of the northern end of the Great Barrier Reef, it is never cold and daytime temperatures in the sun exceed 32°C; most Noddy Terns (*Anous stolidus*) return from the sea when it is almost dark and do not depart until several hours after sunrise (Warham 1961).

Effects of Diet and Foraging Success

For some species, the nature and activity of their food influences the time they leave the roost and return. Just as birds with relatively larger eyes may begin singing at lower light levels, a study of the sequence in first arrival times of birds at British feeders during winter mornings found that those with larger eyes relative to their body size came sooner than species with relatively smaller eyes. Given that in winter birds are eager to begin feeding as soon as they can, arrival time at a feeder may be a good proxy for morning roost departure time. For twenty-four species, the spectrum ran from a mean of 15.4 minutes after first light for the Eurasian Blackbird (*Turdus merula*), to 54.7 minutes after first light for the Eurasian Treecreeper (*Certhia familiaris*). Insect eaters tended to arrive earlier than seedeaters, but the sequence reflected relative eye size rather than diet, at least at that season, since in winter many British insectivores turn to plant matter when insects are scarce. The ease at which food can be discerned and handled at a feeding station may enable some birds to forage earlier there than in natural situations. Other factors influencing the arrival times may be avoidance by some species of aggressive competing species at the feeder, as well as avoidance of nocturnal and crepuscular predators that might be encountered on the way to the feeder (Ockendon et al. 2009). Insect eaters such as bee-eaters and New World flycatchers, which both pursue flying insects they spot while perched, rise later than species that pick food off the ground or foliage. Like swifts, they may need to wait longer for these insects to become active (Thomson 1964, p. 708).

The roosting hours of predatory birds both reflect the daily cycle of their prey and in turn influence the hours their prey are active. In urban and rural Indiana, wintering Cooper's Hawks (*Accipiter cooperii*) and Sharp-shinned Hawks (*A. striatus*) overlap in habitat but less so in hunting hours and actual prey. There, 80 percent of Cooper's Hawks typically leave their roost before sunrise and 60 percent return after sunset; these are their peak hunting times. In urban areas, they focused on feral pigeons (*Columba livia*), Mourning Doves, and European Starlings, which may be easier to catch while these are still roosting. Some 50 percent of Sharp-shins set forth after sunrise, some not until an hour later, and 80 percent return well before sunset, possibly reducing their own risk of predation by owls. Eleven of twenty-three Sharp-shins tracked in this study were killed by owls. Just as the Sharp-shins may be avoiding the light levels most dangerous to themselves, so the activity peak of their winter prey, mostly juncos and sparrows, occurs before sunrise, perhaps to avoid the hawks (Roth and Lima 2007).

For raptors whose food comes in large packages, the time of the last meal may also influence roosting time—unlike small birds that continue feeding until late in the day. The Indiana Sharp-shinned Hawks often returned to their roost after they had made a kill, while unsuccessful birds did not return until the end of the day. Cooper's Hawks did not return until sunset or later, whether or not they had captured something earlier (Roth and Lima 2007). In winter, Rough-legged Hawks (*Buteo lagopus*) arrive at their roost from several hours to immediately before dark, with successful hunters returning earlier (Bechard and Swem 2002).

Summary

Several factors influence the time that birds approach and depart their roost: light levels, weather, distance to foraging sites, hunger, and even the moon, which enables some birds to continue feeding or traveling when it lights the sky. For diurnal birds, light level is probably the most important factor, because this affects their abilities to feed and to find their way to and from their roost. Many birds, especially those moving alone, are also concerned about how visible they themselves may be to predators. Woodpeckers and other cavity roosters often return early to their roost to ensure that no other bird takes it, and they stay there longer in the morning. Some crepuscular and nocturnal birds also time their travel to avoid predators, while, for other nocturnal birds, their schedule is set by the activity periods of their prey. Weather conditions, both the normal range of factors in each season such as cloud cover as well as extremes in temperature, further influence the time birds spend roosting.

Any bird's activities during the twenty-four-hour cycle may vary throughout the annual cycle, based on its age, sex, fitness, and latitude. While light level and day length are closely related, each may independently affect the time spent roosting. In spring and summer at higher latitudes, when days are long, many birds, even when they have young to feed, can finish their day and return to their roost at higher light levels than during fall and winter, when they must use all available daylight to feed only themselves. In the darker seasons, many birds leave their roost at lower morning light levels than during spring and summer, to maximize foraging time during the short days.

Birds are more consistent about when they depart their roost than when they return, because their return time reflects their feeding success and distance traveled. Those that have fed well may return to the roost at higher levels of daylight than birds that have had less success and continued feeding as late as they could. Weather also affects the time birds leave and return to their roost. On cold, dark, and rainy mornings birds often wait to emerge until at least slightly later. Those roosting in cavities often wait longest. Some birds return to their roost early when they anticipate storms. During extreme cold, some spend more of the day at their roost even when it is bright—and some birds do not leave it all day.

8

To and From Communal Roosts

Birds that roost singly usually approach and depart from their roost as discreetly as possible, to avoid calling attention to themselves. Among birds that roost in large aggregations, however, arrivals and departures are more elaborate, often noisy and conspicuous. They may be in several stages or a single flight, with social interaction or solitary purpose, reflecting the functions, benefits, and hazards of the roost.

Pre-roosts

Many of the species that roost in large flocks and disperse widely during the day to feed have rituals ensuring that birds return to an established roost and that newcomers can find it. This includes staging areas where birds gather on their way to the roost, known as pre-roosts. By being closer to where the birds may spend the day feeding, pre-roosts enable them to form flocks at greater distance from their destination and then travel in greater safety.

Locations

Pre-roosts are usually situated prominently, in treetops, open ground, and on buildings. Flocks or individuals at the most distant pre-roosts arrive from different directions. Many species with pre-roosts in leafy trees or in other thick obscuring vegetation are extremely noisy at these gatherings; the sound likely helps approaching birds locate the flock if visibility is poor due to concealment or low light levels. Others that assemble in the open, like Cattle Egrets (*Bubulcus ibis*), are quiet; their exposed location as well as their white plumage, both conspicuous from the air, may be enough to attract the attention of arriving birds. In some species, a few birds remain at the pre-roost when the flock takes off, or some in the flock quickly turn back and land again at the assembly point. This may help ensure that birds arriving slightly later find the site, forming another flock that then travels together. How the remaining or returning birds benefit is unclear. They may have thought the departing flock was too small for safety (Ward and Zahavi 1972).

Pre-roost sites are at varying distances from the roost, reflecting, in part, how far birds disperse every day. Fish Crows (*Corvus ossifragus*) at a pre-roost in trees at the edge of a South Carolina creek fly 8 km to their roost farther down the same creek (Post 1967). Common Ravens (*C. corax*) roosting on transmission towers running between Oregon and Idaho have staging areas several kilometers from the roost. Some ravens leave these pre-roosts at least an hour before sunset, with others birds only arriving at the tower roosts after dark (Engel et al. 1992). The ravens feeding at a composting facility in Thuringia, Germany, have only a five-minute flight to the pre-roost (Janicke and Chakarov 2007). Near Davis, California, the distance of American Crow (*C. brachyrhynchos*) initial pre-roosts to the roost is more than one kilometer, and they range from 200 meters to several kilometers from each other (Verbeek and Caffrey 2002). Rough-legged Hawks (*Buteo lagopus*) wintering in Illinois sometimes gather at a pre-roost 0.4 km away and go to the roost five to fifteen minutes after sunset (Schnell 1969).

Even pre-roosts close to the roost site demonstrate their benefits. Pied Wagtails (*Motacilla alba*) in Britain gather on open ground within a few hundred meters of the roost. They stand there quietly, continuing to feed or preening themselves, with little aggressive interaction. Shortly before dark, when it is safer and most of the birds have arrived at the gathering place, the flock flies on to the actual roost in taller, thicker vegetation, settling in a group near birds that have come from other pre-roosts. Since wagtails roost in more concealing vegetation than is at the pre-roosts, the landing of the flock makes the roost site more visible to other approaching flocks than if birds arrived there singly (Zahavi 1971).

In some cases, however, pre-roosts require a detour. Some Hooded Crows (*Corvus corone*) in Norway were found to travel nearly twice as far from their daytime feeding territory to go to a pre-roost rather than directly to their roost site (Sonerud et al. 2002). Flocks of European Starlings (*Sturnus vulgaris*) in Lancashire, England, that pass their usual roost site on the way to a pre-roost will often circle hesitantly over the roost and then continue on to a pre-roost (Spencer 1966). That these birds chose to spend extra energy at the end of the day to attend a pre-roost when the roost is nearby indicates that the pre-roost has real functional significance.

For other species, the pre-roosts may form a chain, with more distant ones coalescing at closer points and the flocks growing larger as they approach the roost. In Cornwall, Rooks (*Corvus frugilegus*) spend much of their day at and near their rookery, even in winter when the nests are not being used. At that season, the rookery is the first gathering point on the way to the roost, which may be several kilometers distant. From the rookery, the birds fly, mainly along regular flight lines, to fields and trees near the roost. Here, with rooks from other rookeries, they may feed on the ground or move back and forth to the trees. After twenty to thirty minutes, the birds become quiet and all suddenly take off for the roost, which is always in trees that are never used for nesting (Coombs 1976, pp. 102–103).

Just as roosts are traditional and long-standing for many species, so are some pre-roost sites. These may or may not have any distinctive geographic or topographic features to justify their long-term use, and their use may be seasonal or year-round. A seasonal pre-roost site may have persistence even for relatively short-lived birds. In Panama, a roadside tree first noted in 1973 as a place where wintering Prothonotary Warblers (*Protonotaria citrea*) gather at dusk—before flying off in a flock to their roost when it is too dark to follow them—was still used in 1991, several Prothonotary generations later (Warkentin and Morton 1995).

Pre-roost sites are sometimes used by several species. The pre-roosts of Rooks and Carrion Crows (*C. corone*) often also contain Jackdaws (*C. monedula*). These may be going to their own roost and, when flying by, they drop into the area where the other corvids are gathering. Where there are many Jackdaws, they form their own pre-roosts (Coombs 1976, pp. 125, 102). More rarely, the conspicuous pre-roost sites of various corvids are also used by completely unrelated birds. In South Carolina, a pre-roost of Fish Crows at the edge of a swamp was also used by Little Blue Herons (*Egretta caerulea*) and Great Egrets (*Ardea alba*), with each species later flying in its own flight lines toward the area where all three species roosted (Post 1967).

Time of Pre-roosting

The time at which pre-roosts assemble varies among species as well as by season and light level. In South Africa's wet winter, when food for Cattle Egrets is abundant, they begin to gather early in the afternoon; even though the days are short, the egrets cease feeding and congregate in large groups before flying to their roost. During the much longer days of the dry summer, when egrets need more time to forage, they do not assemble until the light levels are lower (Siegfried 1971). American Crows begin gathering in small groups two to three hours before sunset, using trees, buildings, or the ground. From there, they may fly directly to the roost or stop at other pre-roosts on the way, joining other birds (Verbeek and Caffrey 2002). Over the course of the year, the Common Ravens that rely on the unlimited food supply of the composting facility in Thuringia begin arriving at their pre-roost at a sheep farm an average of 108 minutes before sunset, with the final birds coming within thirty minutes after sunset; in winter more birds arrive at lower light levels than in summer. Some ravens continue feeding while in the fields of the farm. More than half the birds then leave for their roost in a woodlot between twenty minutes before sunset and ten minutes after it (Janicke and Charkarov 2007). In Singapore, Common Mynas and White-vented Mynas (*Acridotheres javanicus*) both start forming pre-roosts one to two hours before flying to their actual roost (Nee and Yeo 1993). House Sparrows (*Passer domesticus*) gather at their pre-roost congregating sites 60–106 minutes before sunset and spend 30–60 minutes there, often singing communally; they leave for their roost 15–30 minutes before sunset and, on overcast days, some 15–30 minutes earlier (Lowther and Cink 1992).

For some birds, the pre-roost may begin even earlier in the day. Herring Gulls in central Maine that feed at dumps do not need to spend much time searching for food. After the breeding season, when not caring for young, they can retire to their "club," a place where gulls gather and spend time sitting quietly or sleeping most of the afternoon. As light diminishes at the end of the day, the gulls here stand up, preen, stretch, walk around, and move closer together. After about ten minutes, the more active birds take off toward their roost, followed by others that had been less restless. In another five to ten minutes, the activity begins again, with more birds departing. In each sequence, as darkness grows, more birds fly to the roost. Finally, the last few leave, individually or as a group, when darkness is nearly complete (Schreiber 1967).

Specialized Pre-roost Behaviors

Birds subject to predation when in flight to a roost or at the roost may have additional refinements to their pre-roosting behavior. Small flocks of African Crowned Hornbills (*Tockus alboterminatus*) rotate among six roosts in an irregular sequence. At these roosts, each bird perches on long thin branches or vines with open space all around it so it can escape easily if surprised during the night. To reduce the chance of being discovered at the roost, at dusk the hornbills squat down, as if to sleep, on branches of an apparent roost, and then, when the light is lower, fly off several hundred meters to their actual roost. This may fool predators watching them (Kemp 1995, pp. 32, 110). Southern Ground Hornbills (*Bucorvus cafer*), also African, have another pre-roost behavior that may draw the attention of predators without revealing the actual roost site. The extended family group spends its day on the ground and at the end of the afternoon indulges in a conspicuous bout of play, mainly among the immatures. Play includes bill-wrestling, chasing, and pouncing on one another. Just before dark, the entire family flies into trees and, as darkness falls, makes a long flight to the final roost site (Fry et al. 1988, p. 378).

Dunlins (*Calidris alpina*) on the Alaska coast at the Yukon-Kuskokwim Delta gather at dusk on the flats at the mouth of the Tutakoke River irrespective of the tidal condition, and then at darkness they fly to their roost site; this may reduce the chance of Arctic foxes (*Vulpes lagopus*) finding the roost. It may also ensure that when the birds arrive, having already gathered elsewhere, they are able to assemble in a tight formation, further reducing the probability of detection (Handel and Gill 1992). Other birds that do not move in flocks and therefore are at greater risk while in flight, may have pre-roost sites very close to the actual roost. Gray Kingbirds (*Tyrannus dominicensis*) in Puerto Rico gather on the edge of the mainland before flying 30–35 meters to offshore mangrove islands; they fly low over the water, singly, in pairs, or in groups up to six, starting one minute before sunset, when they are less visible (Post 1982).

Boat-tailed Grackles (*Quiscalus major*) roost communally all through the year, and females sleep on the nest only when incubating and brooding; the pre-roosts are usually separate for each sex, as are the foraging flocks. Birds gather 200–500 m from the actual roost, and they may stay there nearly two hours, vocalizing, preening, bathing, and foraging, until they fly together to the roost a few minutes after sunset. At the male pre-roosts, the birds also sing and display, adding another social function to the assemblage (Post et al. 1996).

Some pre-roosts have only seasonal benefits. During July and August, Clark's Nutcrackers (*Nucifraga columbiana*) in the Tioga Pass, California, gather in late afternoon on the west-facing Mt. Dana side of the pass, where they receive the last sunlight of the day. Between 7:30 and 8 p.m., when light diminishes rapidly, they fly two kilometers across the pass to roost on the east-facing slope of Gaylor Ridge, which is lit by the sun earliest in the morning. The birds may be maximizing their feeding times at both ends of the day as well benefiting from the sun's warmth. By September, when Mt. Dana receives less afternoon light, the nutcrackers cease assembling there (Tomback 1978).

A Complete Sequence in European Starlings

In much of the Northern Hemisphere, the entire sequence of pre-roost movements (where these involve several stages) may be most easily observed in European Starlings. As studied in Illinois farm country, they arrive at the first assembly area from feeding areas in all directions except that of the final roost. These groups fly on to second pre-roosts that are on direct flight lines to the roost; here they may also be joined by other starlings whose initial gathering place was nearby, as well as by birds that came directly from feeding sites. Nearest the actual roost, birds in the various (and now much larger) incoming flocks may scatter to stop at final assembly spots, mixing with others that arrived on different flight lines. From each of these last pre-roosts, the birds go to the roost. These nearby assembly sites will be the last place where starlings can come to the ground and feed, so this link in the chain is more likely at rural roosts than urban ones, to which birds must fly some distance over a built landscape with little or no open land (Davis and Lussenhop 1970).

Weather affects how starlings use pre-roosts. In Lancashire, during afternoons of misty drizzle, small groups assemble at the tops of conspicuous trees. Some groups remain small, departing as the next group

At starling pre-roosts, some birds depart for the subsequent pre-roost while others are still feeding and more are just arriving.

arrives and lands, or they may wait until more accumulate. If there is real fog, the birds wait until it dissipates. In colder weather, starlings may come to the ground and feed while at their final assembly point, immediately before going to the roost. On windy days, flocks going to the roost may not stop at the pre-roosts. If facing into the wind, they fly low and bunched; if the wind is behind them, they fly high enough to be inconspicuous, at least to human eyes, and take on the formation of an open fan (Spencer 1966).

Arrival at Communal Roosts

Some birds that roost communally, even when these sites are conspicuous and traditional and therefore likely known to predators, are, like the birds that roost solitarily or in pairs, also cautious about their approach. They may be more vulnerable during their approach than after they have taken a perch among many others and reduced their odds of being singled out. When after sunset Gray Kingbirds cross singly or in very small groups from mainland Puerto Rico to offshore mangrove islands thirty to thirty-five meters away, they fly low over the water, where they are more difficult to see in the dim light than were they against the sky (Post 1982).

Starlings

Some small, fast-flying birds such as starlings gain safety by approaching their roost in dense flocks that may wheel about while rapidly changing direction and dimensions before settling. The flocks are conspicuous from a distance, but they make it challenging for any hawk or falcon to penetrate or to focus on a single bird. The dynamics within these shape-shifting bodies have been best studied in European Starlings. At any given moment, most birds in the flock are moving in the same direction, in a horizontally thin configuration. The shape of the flock changes when the birds turn, because this compresses the flock so its relative proportions shift. To turn, the birds bank, losing lift while gaining speed; after each bird completes most of the turn, it gains lift, so the entire flock appears to rise in height (Hildenbrandt et al. 2010).

In these starling flocks, the greatest density of birds is at the edges, where this may increase predator confusion. Each bird is keeping track of six to seven neighbors, even as distances between them change when the flock turns. They maintain a greater distance between birds at their front than at their sides, to reduce the risk of collision due to any change of speed. Because starlings have lateral vision, they are better able to see the birds at their sides, typically at least a full wingspan away, and they are not at all able to see birds behind them. When an advancing flock makes, for example, a ninety-degree left turn and moves in the new forward direction, the birds that were at the head of the flock do not maintain their frontal position as would the grill of an automobile. Instead, the starlings that were on the left flank become the lead while the birds that had been at the front become the flock's right flank. By these turns, the birds once at a vulnerable edge are shifted to safer positions that are also less energetically demanding (Ballerini et al. 2008).

Roost Approaches by Other Small Birds

Other birds that roost in dense aggregations have different manners of approach and settling in. After starlings, perhaps the ones easiest to observe in North America are the descents of Chimney Swifts (*Chaetura pelagica*) and Vaux's Swifts (*C. vauxi*), with flocks swirling around, sometimes changing every so often between clockwise and counter-clockwise, depending on the most accessible angle from which to flutter into the chimney or hollow tree. The time required for the entire flock to enter the roost depends on its numbers. At two sites in Texas, it was found that, in spring, northbound flocks of Chimney Swifts took an average of 2.9 minutes after the initial circling for the first birds to enter the chimneys, while some birds circled for thirty minutes before descending. On evenings with strong winds, rain, or thunder, birds entered more rapidly, without circling (Michael and Chao 1973). This author has also seen Chimney Swifts accelerate the process when a Peregrine Falcon (*Falco peregrinus*) flies by.

Flocks of swallows coming to roost, usually using more expansive sites and therefore in vastly greater numbers than swifts, also sometimes circle before descending. The ornithologist who came upon the roost site of at least one million Barn Swallows (*Hirundo rustica*) in South Africa in March 1954 noted that the cloud of birds began above the reed bed at 50–75 m and was dense at least up to 250 m; its width was 400–500 m in diameter. "The flock as a whole was sharply delimited and kept over the same place all the time. . . . The immense stream of Swallows flowing into the main crowd from W.S.W. was also very sharply delimited, and the birds very densely packed; indeed, they formed a dark band, about 100 metres up in the air and 40–50 metres wide. . . . The cloud of Swallows was whirling around rather high up in the air until twilight. The birds even seemed to get more active before alighting. . . . Finally, the birds alighted in the reeds. Some of them dropped suddenly, and then it was a question of seconds until the others followed. Like a torrential shower of rain they hurled themselves down more or less vertically, and disappeared instantly" (Rudebeck 1955).

On a much smaller scale, a roost of five thousand to eight thousand Cave Swallows (*Petrochelidon fulva*) in a cane field in El Salvador began each evening with the birds appearing about thirty minutes before sundown, foraging widely over the field. At sunset, the birds began circling in an ever-tighter wheel that gained altitude with each revolution. After reaching one thousand meters in fifteen minutes, the flock began to drop, spiraling downward in a funnel shape; when birds were about five meters off the ground, they broke away and flew horizontally until settling in the cane. Here individuals were more widely dispersed, each separated by several meters from its neighbors, than in swallow roosts containing many more birds (Komar 1997).

Dickcissels (*Spiza americana*) wintering in Trinidad fly in direct lines from feeding areas to their roost site. At the end of the afternoon, flocks in fields take off and settle again, taking off and wheeling over the feeding ground and finally setting off in groups of two thousand to three thousand. If the roost site is distant, the birds may amalgamate on the way with others. In one case, they formed a continuous column, twenty to thirty birds wide, that stretched 1.6 kilometers. They fly sixty to ninety meters off the ground, sometimes at the speed of 72 km per hour. Arriving above their cane field roost, the flock breaks up, cascading downward. A flock approaching close to ground level will land at the edge of the cane; the birds then work their way through the vegetation or in short flights to the roost area (ffrench 1967).

Parrots

Among the parrots that roost in flocks rather than in families or pairs, the return to the roost at the end of the day ranges from noisy and conspicuous aerial displays of hundreds of birds to extreme caution. William Beebe, working in a Guianan rainforest in 1916, watched small groups of an unidentified parakeet fly silently from different directions in late afternoon to the top of a sixty-meter tree, above the rest of the canopy. When several hundred had assembled, silent and invisible in the foliage, they would all suddenly take off after several minutes with a great clamor and fly above the forest in circles 0.8–1.6 km in diameter. As they passed over a patch of bamboo that was their actual roost, some parakeets would descend toward it, but if the majority of birds continued circling, these few would return to the flock. Eventually, the entire flock spiraled downward to a tree near the bamboo, remaining there for a few minutes, and then all the birds rose and dove into the bamboo for the night (Beebe et al. 1917, pp. 97–98).

In southeastern Brazil, Red-spectacled Amazons (*Amazona pretrei*) approach their actual roost trees from all the directions in which they have dispersed to feed; they wheel once or twice and then land. During the time between their arrival and nightfall, sometimes ninety minutes, large groups of perched birds rise suddenly in the air, usually with no apparent provocation, and make wheeling, screeching flights of several minutes over the general vicinity. When it is nearly dark, the entire flock does the same, noisily circling the roost trees for ten to fifteen minutes before silently settling down (Belton 1984, pp. 538–539). The surviving population of the much rarer Red-tailed Amazon (*A. brasiliensis*), also in southeastern Brazil, roosts in mangroves and swamp forests on off-shore islands. There, rather than circling conspicuously over the trees, the parrots twist and tumble while in fast swooping flights through the trees, screeching constantly, until they settle (Forshaw 1989, p. 610). Similarly, in Australia, Galahs (*Eolophus roseicapillus*) do acrobatics while flying low and fast, singly and in pairs, in and around their roost trees (Warham 1957).

Some parrots, such as the Crimson-fronted Parakeet (*Psittacara finschi*), simply fly in during late afternoon from wherever they have spent the day and assemble with much chattering in trees near their roost. At the Las Cruces Biological Station in Costa Rica, this is a mature oil palm. Pairs of adults groom themselves and each other. If any in the flock, totaling about fifty, are alarmed, the entire group shrieks—the more birds in the group, the more easily the shrieks are provoked. A small group of parakeets will in due course take the lead in flying into the palm and perching on a frond or on one of the hanging fruiting bodies. Other groups follow, with each bird gradually working its way in toward the more sheltered places where, singly or in pairs, they will spend the night (Bond and Diamond 2019, pp. 47–49).

Much more cautious is the highly endangered Lear's Macaws (*Anodorhynchus leari*) of northeastern Brazil, which roosts on ledges and in holes on canyon cliffs, dispersing before dawn to feed. Decades of hunting for the pet trade may have increased the wariness of the surviving remnant population. At one site where, in 1983, a flock of twenty-three lived, their distinctive roosting habits were observed. A few minutes after sunset, two or three individuals returned, calling as they flew over the canyon and landing in the tallest nearby trees, where they sat quietly for about ten minutes. Then, these "scouts" called loudly and the rest of the flock approached, flying over the canyon, screaming, finally landing near the scouts and sitting quietly. By then it was already dark. The birds screamed again as they flew to the cliff edge, landing there and sitting before finally quietly entering their cavities. The "scouts" then flew off to roost in an adjoining canyon (Yamashita 1987).

Ravens

The communal roosts of nonbreeding Common Ravens may be in trees, on cliffs, in abandoned buildings, along transmission lines, or in concealing marsh vegetation. Some roosts are traditional, while, in places where food is widely scattered, the roost sites shift frequently to safe locations nearby. With this variety of sites and time frames for their use, the approaches also vary. At long-standing roosts used primarily by local and migrant birds that are not paired or defending their own territory, there are often soaring displays involving many birds, with groups spiraling high over the roost as some birds descend to find a place while other arriving birds join the soaring flock. There may also be chases and acrobatics. Where the roosts are transient, the soaring displays take place just before the group moves to another site closer to a newly discovered carcass (Marzluff et al. 1996).

In contrast, at a winter raven roost in reedbeds on the edge of Malheur Lake, Oregon, that on at least one night held as many as 836 ravens, the birds approached by flying ten to fifteen meters off the ground in direct lines, finally folding their wings and collapsing in the dense vegetation and becoming invisible. They came from several pre-roosts within one kilometer, each on open ground where the birds were conspicuous from the air. At these pre-roosts, some birds fed, begged, or chased others, with some pairs soaring above. These gatherings may have served the same locating function that larger numbers soaring over the actual roost does elsewhere (Stiehl 1981).

Mixed-Species Roosts

Four species of herons using a communal roost in New Jersey each approach it at different elevations. Little Blue Herons and Snowy Egrets (*Egretta thula*) fly in at thirty meters then suddenly make twisting dives at sharp angles into the trees. Great Egrets, larger and likely less at risk, usually glide in at less than fifteen meters, often approaching the roost at tree-top level. Green Herons (*Butorides virescens*), much smaller, fly in directly, usually at fifteen meters or less (Seibert 1951).

At multispecies blackbird roosts, Bobolinks (*Dolichonyx oryzivorus*) approach at the highest elevations and in the smallest flocks. Next in order of height are starlings, grackles, and Brown-headed Cowbirds (*Molothrus ater*), with Red-winged Blackbirds (*Agelaius phoeniceus*) closest to the ground. While some red-wings that began sixteen kilometers from the roost fly toward it at 300–450 meters over the ground, the last birds of all species that reach the roost at dusk are just skimming over the vegetation where, like the Gray Kingbirds over the water, they are hardest to see or to pursue (Meanley 1965).

Dreads

A distinctive feature of some birds that roost in flocks where they can easily take off in dim light, as from the ground, marsh vegetation, leafless trees, or buildings, is what are called dreads, or panics, in which many or all the roosting birds fly up, circle around once or several times, and return to their places—all with no evident provocation. While this has not been tested experimentally, dreads are believed, at least for some species, to be an expression of migratory restlessness. They are most conspicuous in spring and autumn. Dreads may take place in the first period of near darkness, or all through the night, perhaps affected by the extent of moonlight, and again before dawn.

At the gigantic southern mixed-species roosts of blackbirds and starlings, dreads occur in late winter, just before the roosts begin to break up or diminish in size. At a 5.6 hectare wooded plot in the Arkansas Grand Prairie, the entire roost population, estimated at twenty million birds, will fly up, wheel around, and return—sometimes several times in an evening—before the birds settle down (Meanley 1965). In British starling roosts, the dread flights are seasonal but independent of temperature. The dreads begin an hour after the starlings first settle; they may continue for four or five hours, starting again before dawn. They typically occur at intervals of four minutes, when a part of the roost takes flight for no apparent reason and lands again after a few minutes. Four minutes is also the usual interval at which in the morning large numbers of starlings take off from the roost to disperse and feed (Spencer 1966).

Terns of several species do something similar while at their nesting colonies and perhaps at other seasons when large numbers come to roost on sandbars; the winter behavior of most highly migratory terns is still too poorly studied to state this with certainty. At dense and noisy colonies, where the birds may be almost as closely packed as at roosts, terns will suddenly grow silent, take off en masse low over the water, and circle back to their nests, with no obvious provocation such as a predator overhead. In Norfolk, England, this has been observed in colonies of Common Terns (*Sterna hirundo*), Roseate Terns (*S. dougallii*), and Sandwich Terns (*Thalasseus sandvicensis*), but not in the smaller, more scattered colonies of Least Terns (*S. albifrons*). Each dread lasted only seven to twenty-five seconds. Between 6 and 8 p.m., there were as many as twenty-six dreads at a colony of Common Terns (Marples and Marples, p. 171). In Norfolk and also in Common Tern colonies in Ontario and New York, dreads occur most frequently early in the nesting season and were only observed during daylight (Morris and Wiggins 1986; Burger and Gochfeld 1991b, p. 191), but this author has seen them at another New York colony at dusk in August, after young birds have fledged. End-of-season dreads may reflect some of the premigratory restlessness hypothesized for the starling and blackbird dreads at their roosts.

Leaving Roosts

Birds that roost alone generally leave their roosts quietly in the morning, whether they depart at low light levels or wait longer. The departures of flocks from their communal roosts are inevitably more conspicuous and sometimes deliberately amplified with sound. But whether the flocks are early or late risers, their dispersal is usually more rapid and direct than the gathering at the end of the day, because birds are eager to begin feeding. Some also want to drink before they undertake any other activity. And, since most birds that roost in flocks have come from some distance, to which they will likely return in the morning, it will be longer before they start feeding again than for any bird that roosts solitarily within its own territory.

As with the roost arrivals, departures have been most extensively studied in European Starlings. At British roosts, starlings leave in a series of waves. The numbers leaving in each wave vary, while the intervals between them average 4.5 minutes, with a range of half a minute to twenty-two minutes, and with the least regularity in summer. The number of waves ranges from one to ten, with more at the high end in summer than in colder seasons when birds may be more eager to begin feeding. At no season, however, does the number of waves correlate with the size of the roost or the area from which it draws its members. At least a quarter hour before the first flight, starlings begin vocalizing, building to a crescendo when suddenly, as some birds take off, there is a hush from both the flying birds and those that remain. The noise resumes when the last birds in each wave have left the roost. Some birds that were in the wave may circle back, just as some do when a flock leaves a pre-roost, perhaps to join the subsequent wave (Spencer 1966). More recent research using radar has found that the flight speed of starlings leaving British roosts is greater in winter than in summer (74 and 59 km/h, respectively) and that at any season flight speed is greater in earlier departure waves than later ones. Birds in the early waves are also heavier than those in the subsequent ones, mostly adult males, while younger females depart later (Feare 1984, p. 244).

At the one-million-plus roost of Barn Swallows in South Africa, the birds similarly left in waves. The first, at 5:50 a.m., gave the impression of an explosion, with an estimated 500,000 birds "thrown up high in the air, as if forced to do so because there was no space left for them immediately over the reeds," where a compact mass of swallows was thronging. The birds then flew a few meters above the ground, all heading west or southwest. Four minutes later, another half million flew up, although not with the same explosive aspect, and flew low over the ground in the same direction as the first flock. A third wave, of 100,000, some "thrown up" 40–50 meters, took off five minutes later, and two minutes after these, another 100,000 rose and departed in the same manner. By 6:03 a.m., the last few thousand swallows left the reed bed, as smaller flocks of the several weavers also roosting there flew about in different directions (Rudebeck 1955).

Dickcissels roosting in sugar cane in Venezuela leave their roost more rapidly and more conspicuously than they arrive. They begin ten to thirty minutes after sunrise, usually emptying the roost in less than thirty minutes. At some sites that include nearly three million birds, hundreds of thousands rise at once, spiraling upward in tornado-shaped funnels. Other birds leave in broad serpentine columns with the leading birds above those behind. At smaller roosts of less than 100,000, all the birds may leave at once, but large roosts have several departure waves, sometimes simultaneous from different parts of the roost going in different directions. At dawn, sometimes small groups of fewer than one hundred are just arriving at the roost, before any birds there have departed; they presumably were not able to each the roost the evening before and return now to rejoin the larger flocks (Basili and Temple 1999).

The morning departure of parrots from their roost has the same range of behaviors as their evening arrivals. The ultracautious Lear's Macaw scouts, which first inspected the flock's roost site before leading the other birds there, emerge before sunrise from their own roost in the adjacent canyon and fly screaming over the canyon that has the flock's nest holes. These birds then fly out of their holes and the entire group flies out of sight (Yamashita 1987). The Crimson-fronted Parakeets in Costa Rica roosting in palms begin vocalizing just after dawn; they begin with low, purring calls and gradually switch to the louder flight call. As more birds join in, small groups begin flying out of the palm while calling intensely. Within an hour of first light, few parakeets remain (Bond and Diamond 2019, p. 47). Yellow-shouldered Amazons (*Amazona barbadensis*) in Bonaire, Netherlands Antilles, start calling shortly before sunrise; soon the entire flock is screeching. After several minutes, they leave their perches, flying to nearby trees and cacti, moving about restlessly from one to another for fifteen minutes, and then fly off to their feeding areas. In Australia, several species of cockatoos and corellas leave their roosts in trees before sunrise and fly to nearby water to drink. The Palm Cockatoo (*Probosciger aterrimus*) of New Guinea roosts singly on the uppermost bare branches of tall trees at the edge of rainforest; only well after sunrise do they begin preening and then gather in small groups in large trees where they bow and display before going off to feed (Forshaw 1989, pp. 618, 140–148, 119).

Other large forest birds, such as hornbills, also move from their roost to a prominent perch where they stretch and preen. This may be followed by bouts of territorial calling and display before they begin feeding. Silvery-cheeked Hornbills (*Ceratogymna brevis*) of eastern Africa roost communally in groups of up to two hundred birds; before they leave the roost, they bask in the sun, with head lolling about. About thirty minutes after sunrise, they disperse to feed, singly, in pairs, or in small groups (Kemp 1995, pp. 24, 258). Great Blue Turacos (*Corythaeola cristata*) in forests of west and central Africa roost in tall trees rising over the canopy; the small flocks that roost together leave the tree one after another, flying silently to a nearby tree where they call loudly and then fly again to more distant feeding trees (Forshaw 2002, p. 53).

For flock-roosting species that are very noisy before departure, the volume may be an aspect of communication preparing the birds to leave. Flocks of Jackdaws begin calling before dawn, with the vocalizing continuing for an hour or more, increasing in volume. A study of Jackdaws at several roosts in Cornwall, UK, found that the steeper the rise in intensity of the calls, the sooner the flock departed, and the more likely the flock would take off in unison. On twenty-one of thirty-three mornings, all the birds flew up within five seconds. To test whether a certain intensity of calls was required before the Jackdaws left their roost, recordings of roosting calls were played from several speakers within a roost to generate an earlier onset and peak of calling. This accelerated their readiness to depart, resulting in the first mass departure nearly seven minutes earlier than when no calls were played or when recordings of wind were played. The vocalizations and their intensity seem to facilitate cohesion within the flock, signaling willingness to leave. The benefits of this cohesion are, as in all flock departures, reduced risk of predation and, perhaps for some birds, guidance to known feeding areas (Dibnah et al. 2022).

Post-roosts

The distinction between flying a short distance from a roost to stretch, preen, bask, and socialize before going off to feed, as do some parrots and hornbills, and making longer flights with roost-mates where they pause for more than a few minutes before dispersing to the day's destination may be a fine one. The latter behavior is known from far fewer species, albeit across a broad ecological spectrum including ground feeders, scavengers, and aerialists. Some birds with post-roosts feed at patchy sources of food and others at foods that are evenly dispersed. These post-roosts may for some birds be the opportunity to gain information on current good feeding sites, while for others they may simply be sites where they can warm up more efficiently than at their roost.

On the Dutch coast, wintering Barnacle Geese (*Branta leucopsis*) roost at night in salt marshes and feed during the day in nearby pastures. Each morning, they first fly in small groups to nearby mudflats, where they may spend almost an hour—10 percent of the available daylight— when the pastures are only a five-minute flight away. The geese spend longest in the morning on the mudflats following evenings when some birds returned late to the roost site, indicating they needed more time to feed, wherever they were. The individuals that depart soonest from these post-roosts may be the ones most confident of finding or return-ing to a good feeding area. Whether such gatherings are actually a way for less knowledgeable members of the post-roost to gain information has not been tested (Ydenberg et al. 1983).

The post-roost may for some birds simply be a way to wait until they are warm enough, until flying conditions are right, or until enough birds are ready to begin feeding. On cold, frosty mornings in South Africa, Cattle Egrets departing from their roost may either go directly to feed-ing grounds or may stop somewhere on the way to perch in the tops of trees that catch the early morning sun. But this behavior is not restricted to cold days—it does not take place every cold day and it sometimes occurs on warm mornings as well (Siegfried 1971). Turkey Vultures leave their roost for a post-roost site, sometimes one kilometer away, if it provides better opportunities for sunning or ease of takeoff when air currents begin later in the day; this may be as many five hours after sunrise or, on rainy days, not at all, in which case vultures may remain at the post-roost all day (Kirk and Mossman 1998).

During the winter, British roosts of Pied Wagtails may hold several hundred birds. They leave the roost some thirty minutes before sunrise for a post-roost. Near Reading, wagtails gather on the roof of a factory about fifty meters from the roost; counted over several mornings, this site was found to attract a mean of 13 percent of the birds at the roost. The first arrived in semidarkness, later on overcast days, with the peak arrival time coinciding with the first departures of birds in different di-rections to their own feeding areas. While at the post-roost, most birds sat in rows on top of the glass windows of the roof, often close to warm ventilator shafts (Broom et al. 1976). So, for the wagtails, like the Cattle Egrets and the Turkey Vultures, the post-roost may be an opportunity to warm up before beginning the more energetically demanding activi-ties of the day.

Among sedentary starlings that maintain their territories throughout the year and do not join communal roosts, each bird, and sometimes two, but not necessarily a pair, spends the night in a tree hole. They emerge in winter as many as forty minutes before sunrise, while in spring and autumn usually fifteen minutes before. The first to emerge in the morning flies to a nearby tree, settling in its inner branches; it is soon joined by other local starlings, usually the same number at each post-roost, and in the same tree. In one area in Britain, the number at these gatherings ranged consistently from seven to twenty-four. The birds preen and sing or sit idle for about twenty minutes and then disperse in ones and twos to feed, sometimes first returning to their roost hole (Spencer 1966).

Perhaps similarly, during the winter in Lahore, Pakistan, a few House Swifts (*Apus affinis*) leave their nests even before sunrise and, well before most others, fly high over their colony, often stopping in midair to hover. As more birds join them later in the morning, the amount of hovering diminishes, but circling over the colony continues until about 10 a.m., when the birds disperse beyond the city limits to feed and are not seen again at the colony or anywhere over the city until they return at the end of the day (Razack and Naik 1965). For birds as aerial as swifts, this stationary gathering of a few hours each morning could also be considered a post-roost.

For other birds, hesitancy to be alone at an exposed feeding site may make birds stay in a group akin to a post-roost even later in the day. White Ibis (*Eudocimus albus*) in South Carolina gather after their initial morning feeding for extended periods in trees at the edge of salt marshes, far from their nocturnal roosts. Since their diet consists of hard-shelled crabs, these midday roosts may be for digestion. A single bird leaving the daytime roost and landing in the marsh will turn and stand, looking back at the others without feeding until more birds join it; if others do not arrive soon after, it will return to the roost. Sometimes, single birds, even when still in flight, turn their head to look back at their former companions and, if none are following, will fly back to the roost without landing (Bildstein 1993, pp. 105–106).

Summary

While birds that roost singly or in small groups usually approach and depart from their roost as discreetly as possible (to avoid calling attention to themselves), other birds reduce the danger by arriving at roosts in flocks that may number dozens, hundreds, thousands, or even millions. Their arrivals and departures are more elaborate, often noisy and conspicuous. This may make the roost site, sometimes used for generations, well known to predators, but the dense flocks, often flying about in noisy gyrations before dropping into vegetation, make it hard for aerial predators like falcons to single out and pursue a target. The noise and sight of the birds flying over the roost, as well as the incoming streams of birds that gathered at staging areas called pre-roosts on the way, help guide others that may be unfamiliar with the site, thereby further increasing the numbers. Often, their descent to the roost is highly synchronized, reducing the time they are vulnerable.

The arrival time at the communal roost may vary over the course of the year. When the days are long, some birds gather at the roost hours before dark. During shorter days, they may arrive after sunset, having maximized their feeding time. Other birds, especially those coming singly or in small groups, usually arrive in semidarkness to avoid predators.

Birds approaching communal roosts may move in several stages or a single flight. By being closer to where birds may spend the day feeding, pre-roosts enable them to form flocks at greater distance from their destination and then travel in greater safety. Pre-roosts are usually situated prominently, in treetops, open ground, and on buildings. Flocks or individuals at the most distant pre-roosts arrive from different directions. Many species with pre-roosts in leafy trees or in other thick obscuring vegetation are extremely noisy at these gatherings; the sound likely helps approaching birds locate the flock if visibility is poor due to concealment or low light levels. Others assemble in open areas where they are conspicuous.

At the roost, some birds perform what are called dreads or panics, in which many or all the roosting birds fly up, circle around once or several times, and return to their places—all with no evident provocation. While this has not been tested experimentally, dreads are believed, at least for some species, to be an expression of migratory restlessness. They are most conspicuous in spring and autumn. Dreads may take place in the first period of near darkness, or all through the night, perhaps affected by the extent of moonlight, and again before dawn. For some birds, these flights occur at the same intervals as flocks depart in the morning from a large communal roost.

The departures of flocks are also conspicuous, sometimes deliberately amplified with sound. But as with solitary roosters, their dispersal is usually more rapid and direct than the gathering at the end of the day, because birds are eager to begin feeding. Some also want to drink before they undertake any other activity. And, since most birds that roost in flocks have come from some distance, to which they will likely return in the morning, it will be longer before they start feeding again than for any bird that roosts solitarily within its own territory. For flock-roosting species that are very noisy before departure, the volume may be an aspect of communication preparing the birds to leave. The vocalizations and their intensity seem to facilitate cohesion within the flock, signaling readiness to take off. The benefits of this cohesion are, as in all flock departures, reduced risk of predation and, perhaps for some birds, guidance to known feeding areas

Some birds, in less of a rush, fly a short distance from a roost to stretch, preen, bask, or socialize before dispersing to the day's destination. They may hang about for an hour or more. Birds that do this include ground feeders, scavengers, and aerialists. These post-roosts may for some birds be the opportunity to gain information on current good feeding sites, while for others they may simply be places where they can warm up more efficiently than at their roost.

9

Predation, Parasitism, and Competition

Except perhaps in infancy, birds are never more vulnerable than when they sleep, so it is no wonder they have evolved such a wide variety of means to reduce the risk of predation during these hours. The very nature of their sleep, with its distinctive pattern of extremely short bouts punctuated by brief awakenings, and the ability to maintain alertness with half their brain and one eye while the other hemisphere rests, are testament to the fundamental need for vigilance. To these physiological adaptations, birds have added many behavioral ones, ranging from seeking concealment, rotating among roosts, and isolating themselves in places beyond the reach of predators, to building defensive structures such as dormitories and trying to deceive predators about the roost site. Some birds roost near more menacing species, while others favor sites that have visual barriers, so no predator can approach unseen. Still other birds seek to overwhelm a predator with numbers, sometimes boldly advertising their communal roost with aerial displays and noise that signal to more birds that they should join.

The roosting strategies that are most effective at reducing predation, however, enhance birds' exposure to parasites. Balancing against the safety of cavities comes their harboring parasites, which may be a reason many cavity roosters shift frequently among several sites. Similarly, the birds that huddle together or roost in dense flocks also increase their exposure to parasites as well as disease.

For whichever strategy any species or individual adopts, the sites providing the sought-after benefits may be scarce, be they tree cavities, beaches above the high-tide line, or dense wetlands. This leads to competition among individuals and between species. Such scarcity is reflected in the tendency of tree hole roosters to retire earlier than most other birds, in order to secure their preferred cavity, and in the hierarchies at communal roosts that typically force younger or weaker individuals into more exposed positions. In mixed-species roosts, smaller or less aggressive species often get the perches closest to the ground or most difficult to escape from. Where several species seek a limited resource such as tree cavities, there may be battles among them, as well as competition, if not combat, with other animals, from colonial insects to primates.

Predation

The actual extent of nocturnal predation at individual or communal roosts is challenging to measure. Its occurrence—though not its scale—is somewhat revealed by losses of birds on their nest, remains of consumed carcasses, the stomach contents of and pellets cast by avian predators, observations of predators in action, and recordings by cameras. Owls and diurnal raptors, as well as foxes, domestic and feral cats, rats, mongooses, weasels, and snakes, are all pursuing birds both awake and asleep, so the share of sleeping birds in their diets will be less than their total consumption. Observations of systematic rather than incidental roost predation are most available for species that scientists may be studying for other purposes.

Owls

Small birds living in vast flocks, such as Red-billed Queleas (*Quelea quelea*) in Africa or Dickcissels (*Spiza americana*) wintering in South America, are magnets for predators, by day and by night. At a long-studied site in Nigeria near Lake Chad, Barn Owls (*Tyto alba*) regularly took queleas when these roosted in aquatic grasses; when the flocks slept in thorn trees, as they prefer, Barn Owls did not approach them. While the queleas were roosting in grass, pellets found under the nest of one Barn Owl pair contained about two-thirds quelea skulls, with the balance being remains of rodents and large shrews in equal numbers. Three weeks after the owl nest was found, the queleas abandoned the grass roost, moving to a nearby grove of acacias; from then on, the owl pellets contained only mammal fragments (Ward 1965).

At Dickcissel roosts in sugar cane in Trinidad, Barn Owls arrive at the roost after sunset and may stay for several hours, at least until midnight, and then return before sunrise. They fly low over the cane fields, plunging into a cluster of roosting birds, provoking versions of the Dickcissel's roosting call; the Dickcissels continue with their alarm call for several minutes after an owl has moved on. The rate of roosting calls from these disturbances, every minute or so for long periods, suggests that the owls persist for hours and are frequently unsuccessful (ffrench 1967).

In Puerto Rico, both Barn and Short-eared Owls (*Asio flammeus*) fly over water to attack the blackbirds in mixed-species roosts on offshore mangrove islands (Post 1982). At Bolinas Lagoon, on the southern end of Point Reyes, California, wintering shorebirds are significant elements of the diet of Short-eared Owls, Long-eared Owls (*Asio otus*), and Great Horned Owls (*Bubo virginianus*), the last two being more strictly nocturnal hunters. Given their prime hunting hours, all three owl species are likely to take a significant share of shorebirds when these are roosting. Based on the number of pellets found with shorebird remains during the winter of 1972–73, researchers at Bolinas calculated that the owls and some diurnal raptors altogether consumed 20.7 percent of the Dunlins (*Calidris alpina*), 11.9 percent of the Least Sandpipers (*C. minutilla*), 7.5 percent of the Western Sandpipers (*C. mauri*), and 13.5 percent of the Sanderlings (*C. alba*) wintering at the lagoon (Page and Whitacre 1975).

In some places, owl predation affects the daily movements and roost choices of shorebirds. At Egegik Bay on the Alaska Peninsula, flocks of shorebirds that roost during the day on nearshore intertidal flats do not use them at night, when Short-eared Owls hunt there. At night, the shorebirds move to feed and roost at more distant flats. In addition to the circumstantial evidence that owls may provoke this move, the remains on the beach of a Short-eared Owl, in turn killed by a Gyrfalcon (*Falco rusticolus*), had in its gizzard a freshly consumed Dunlin (Piersma et al. 2006). On the coast of Argentina, wintering Red Knots (*C. canutus*) move every night to feed and roost at more distant sandflats than they use during the day. Both Short-eared and Great Horned Owls are active predators there along the shoreline. No definitive proof has been found that they consume knots, but no other reasons are as plausible for the knots' abandonment of rich daytime feeding sites close to the bluffs where owls may wait at night (Sitters et al. 2001).

At a winter roost of 250,000 American Robins (*Turdus migratorius*) in Arkansas, Great Horned and Barred Owls (*Strix varia*) were continual predators, indifferent to parties of human observers nearby. Barred Owls were heard making their wild calls after killing a robin. Despite the mortality and disturbance, the robins did not abandon the roost (Black 1932).

Diurnal Raptors and Other Birds

ATTACKS ON PASSERINES

A wide range of diurnal raptors also preys on birds at their communal roosts, with the raptors adjusting their foraging timetable to maximize their opportunities. At the Dickcissel roosts in Trinidad, at least one wintering Merlin (*Falco columbarius*) was present every evening that each roost was observed. Two Merlins were frequent and, on some evenings, there were five. They were never present at other hours. Pellets underneath the Merlins' feeding trees contained nothing except Dickcissel remains. On a few evenings, an Aplomado Falcon (*F. femoralis*) was also present, pursing both Dickcissels and bats. When Merlins dived into the cane where hundreds of birds roosted, the Dickcissels scattered and the falcon emerged with no victim. Merlins were more successful when flying a few feet over the ground following a track through the cane, catching Dickcissels crossing from one patch to another. While Barn Owls provoked the roosting call, Dickcissels were silent as long as a Merlin was in view (ffrench 1967). At the Dickcissel roosts in Venezuela, fourteen avian predators preyed on them, including five falcon species, eight hawks and kites, and Barn Owls. Merlins were the most common, with one to five at every roost; in March and April, when wintering Peregrine Falcons (*F. peregrinus*) were moving north, as many as nineteen hunted at one large roost (Basili and Temple 1999).

In southeastern Nigeria, a roost of 1.5 million wintering Barn Swallows (*Hirundo rustica*) using a sloping field of elephant grass (*Pennisetum*) surrounded by rainforest is a comparable magnet for resident and wintering raptors. In early 2001, five species of hawks and kites, two falcons, three owls, and one cuckoo, the Senegal Coucal (*Centropus senegalensis*), were regular predators. All except the African Hobbies (*Falco cuvierii*) hunted by flying low over the roost site, with the diurnal raptors trying to flush roosting swallows, and the owls and coucal catching ones near gaps in the vegetation. The hobbies, sometimes as many as seven, regularly arrived at the site in both afternoon and morning, two to fourteen minutes before the first appearance or emergence of the swallows. Their hunting success rate was 38 percent, highest for hobbies targeting small groups of swallows.

The bulk of the swallows avoid capture by arriving and departing in massive flocks. In the evening, the first swallows appear an average of twenty minutes before sunset, but 92 percent come in the first nine minutes after sunset. The swallows arrive flying very high, forming a dense mass that ascends to at least 500 meters and then dives into the roost at great speed like an inverted tornado, spreading out as they disperse in the tall grass. In the morning, the process is reversed, with 90 percent of the swallows departing in less than ten minutes; some fly downhill low over the elephant grass, while 95 percent fly straight up as fast and as steeply as possible and then disperse from the area at full speed.

The hunting strategies of the hobbies reflect these entrances and exits. They sometimes hunted in twos and threes, with greater success than singly. In the evenings, the hobbies flew high, and the swallows responded by gaining height and swarming in dense masses before they plunged into the grass. In the morning, the hobbies were low, pursuing or diving at the small flocks that flew over the grass, causing the swallows to scatter. For the swallows that left the roost in a steep ascent, the hobbies attacked by stooping from higher up, with an upward swoop at the end of each pursuit (Bijlsma and van den Brink 2005).

One January evening at a roost of more than one million European Starlings (*Sturnus vulgaris*) in Lincolnshire, England, two Northern Harriers (*Circus cyaneus*), a Rough-legged Hawk (*Buteo lagopus*), a

Sparrowhawk (*Accipiter nisus*), two Common Kestrels (*Falco tinnunculus*), a Peregrine Falcon, and two Short-eared Owls—each species usually characterized by very distinctive hunting methods in other situations—were seen pursuing the starlings. At this traditional roost, the pre-roost sites are known to vary, thereby reducing the number of places raptors learn are reliable hunting grounds, while the consolidation of starling numbers at the pre-roosts and their rapid descent at the roost reduce each bird's vulnerability (Feare 1984, p. 236).

Ornithologists in Rome compared the behavior of incoming starling flocks at two roosts that had different amounts of predation by Peregrine Falcons. One roost held fifty thousand birds, the other twenty thousand, and more Peregrines pursued starlings at the larger roost. When attacked, some flocks responded by splitting into two or more parts, often immediately merging again; this could happen several times during a single attack. In other flocks, there were "agitation waves"—changes in density within the flock that took only one second to transmit from one side of the flock to the other. Starlings coming to a roost with Peregrines visible in flight nearby tended to be in tighter flocks. These flocks themselves were visible some ten kilometers away, where flocks going to another roost responded by adopting similar tight configurations. If no Peregrines were active within sight, starling flocks tended to be in a looser formation. Peregrine hunting success was greatest when they could pursue loose flocks or birds flying alone (Carere et al. 2009).

ATTACKS ON SHOREBIRDS

Just as falcons learn to visit roost sites at dusk and dawn, they also can adjust their foraging schedule to the cycle of tides that force coastal shorebirds to roost in dense flocks where only a few places of higher ground along the shore remain above the waterline. In the upper Bay of Fundy, in New Brunswick, as many as 100,000 Semipalmated Sandpipers (*Calidris pusilla*) gather in a single roost during late July to mid-August, when they stage in the bay before flying to South America. At one site, as many as twelve Peregrine Falcons sometimes come to prey on them.

The peregrines used stealth attacks, flying low over the adjacent salt marsh. During a five-year study, the attacks were usually one to three per hour during the period the sandpipers were waiting for the tide to recede. Peregrines did not launch attacks as soon as the sandpipers arrived, but 75 percent of attacks were in the first twenty-five minutes of their presence, with a few in the later stages as well. Early attacks may reduce the likelihood that a Peregrine waiting in trees beyond the salt marsh is perceived, and the attacks may take place before experience leads sandpipers to increase their general vigilance. Larger flocks were less responsive to attacks than small ones, but even the smaller flocks that flew off returned to their roost site, indicating that few alternatives were available nearby (Beauchamp 2016).

At the Banc d'Arguin on the coast of Mauritania, a major wintering site for shorebirds breeding in Eurasia and Greenland, ornithologists found three Lanner Falcons (*Falco biarmicus*), two Peregrine Falcons, and two Barbary Falcons (*F. pelegrinoides*) with completely overlapping hunting areas in one small area during October 1988, when many migrants had only recently arrived and some were in poor condition after their long flights. Marsh Harriers (*Circus aeruginosus*) and Short-eared Owls are also known as occasional predators of shorebirds there. The falcons all hunted at the tidal roosting sites, with their hunting peak in the two hours before high tide, tailing off rapidly afterward. The only successful hunts seen were around high tide, with each falcon hunting for one to two hours per tidal cycle, even when there was only one high tide during daylight hours. Between hunting bouts, each falcon rested or preened for short periods. Later, the prevalence of falcons resting with bulging crops indicated their facility at catching shorebirds.

The falcons there used four hunting methods. Of twenty-two observed low-level surprise attacks from low dunes, like the ambushes in the Bay of Fundy, four were successful. Other low-level attacks flushed entire roosting groups, which flew in dense twisting and turning flocks; the falcons stooped on these, pursuing any bird they could isolate. Of ten such attacks with known outcomes, three were successful, requiring two to six stoops. Only once was a stoop from a falcon soaring at great height witnessed; it was not successful. Finally, the resident pair of Lanner Falcons sometimes hunted together, attacking flocks and individuals seemingly at random. In only one of five hunts, each lasting several minutes, did the female catch a Dunlin. Dunlins were, in fact, the most frequently caught species of the fifteen wintering shorebirds. Both in the Dunlins and the other species, juveniles were caught more frequently than adults. The cumulative average success rate for all hunting methods was 18.6 percent (Bijlsma 1990).

Large daytime roosts of shorebirds also attract other avian predators. Dunlins resting between tidal cycles at the mouth of the Tutakoke River on the Yukon-Kuskokwim Delta of Alaska during late August and early September are frequently disturbed by several nonraptor predators, including Parasitic Jaegers (*Stercorarius parasiticus*), Long-tailed Jaegers (*S. longicaudus*), Glaucous Gulls (*Larus hyperboreus*), and Mew Gulls (*L. canus*), as well as by Northern Harriers and Short-eared Owls (Handel and Gill 1992).

Other Predators

Non-avian predators, especially nocturnal ones that routinely hunt birds that roost alone, are rarely observed, and their impact is difficult to quantify. Small birds sleeping on open nests are an especially easy target for mammals and snakes, because they cannot take a different roosting site every night and because they are more reluctant to fly. The heightened risk of predation for birds sleeping at their nest may be one reason many small birds cease spending the night on their nest as soon as the young can regulate their own temperature.

MAMMALS

The same massive roosts that attract many avian predators also attract terrestrial ones. These, however, are usually harder to detect, especially in thick vegetation during darkness, so data are fewer. At the Venezuelan Dickcissel roosts, three mammal species—one felid and two mustelids— were observed preying on birds: jagarundi (*Felis yagouaroundi*), grison (*Galictis vittata*), and tayra (*Eira barbara*) (Basili and Temple 1999). At the Nigerian roosts of Red-billed Queleas, the only suspected mammalian predators were Gambian mongoose (*Mungos gambianus*) and serval (*Felis serval*) (Ward 1965). In the Arkansas roost of American Robins, bobcats (*Lynx rufus*) and feral house cats (*Felis catus*) were common predators, with as many as four bobcats visible at once (Black 1932). On the Yukon-Kuskokwim Delta of Alaska, arctic foxes (*Vulpes lagopus*) prey on flocks of Dunlins at their roosts (Handel and Gill 1992).

In Africa, Green Woodhoopoes (*Phoeniculus purpureus*) roost throughout the year in family groups in tree cavities or under bark. Any animal that can reach these sites and prevent the birds from escaping therefore has a bonanza. Among arboreal mammals, their predators are the African wild cat (*Felis libyca*) and feral domestic cat, large-spotted genet (*Genetta tigrina*), and slender mongoose (*Herpestes sanguineus*). All these mammals may also be searching for other birds roosting in the trees as well as for the abundant acacia rats (*Thallomys paedulcus*), which are also active at night. Woodhoopoes have evolved a distinctive response that may deter some predators; when confronted in their roost, they secrete a drop of a malodorous liquid from their oil gland (Ligon and Ligon 1978).

SNAKES

Nocturnal snakes take their share of small sleeping birds, but few studies have quantified their impact. In the tropics, where there are many more snakes than in temperate regions, and more of them are arboreal, their toll on roosting birds is likely greater. A seven-year project in central Texas using video cameras at 133 nests of Golden-cheeked Warblers (*Setophaga chrysoparia*) at two locales found that twenty-one nests were depredated by rat snakes, which hunt mainly at night. Ten Texas rat snakes (*Elaphe obsoleta lindheimeri*) and one Great Plains rat snake (*E. guttata emoryi*) consumed the contents of twenty nests and one nest, respectively. At six of the depredations, the incubating female was also taken, five of these by Texas rat snakes. Twenty of the twenty-one incidents were at night, in one area between 8:01

to 11:52 p.m. and at the other between 12:22 to 4:48 a.m. Four incidents were in the incubation stage and seventeen in the nestling stage. Of the latter, four occurred when the young were less than six days old and were still being brooded; there, the female was taken as well. The balance took place when the females were no longer brooding (Reidy et al. 2009).

In a companion study on Black-capped Vireos (*Vireo atricapillus*) that nest in the same region, Texas rat snakes did not capture any adults on the nest—only eggs and young. This may be because the shrub limbs from which vireos suspend their nests are less sturdy or stable than the tree branches warblers use, so the adult vireo is alerted in time by the disturbance (Reidy et al. 2009).

AVOIDANCE STRATEGIES

Some birds adjust their roosting habits in response to mammal predation. At a colony of Black-headed Gulls (*Larus ridibundus*) in Cumberland, England, where red foxes (*Vulpes vulpes*) take many birds, especially at night, the gulls roost in the open on beaches, where foxes normally cannot approach undetected. Only gulls laying and incubating eggs spend the night on their nest. Later in the season, parents leave the colony at dusk to roost on the beach, some leaving behind vulnerable young (Ashmole 1963).

On Ascension Island at 11°S in the mid-Atlantic, feral cats have been common on the island since 1815, soon after British occupation. Due to their predation, Sooty Terns (*Onychoprion fuscatus*), which used to breed all over the island, are now restricted to a few locales. Before beginning to nest, they gather to roost at the start of the season in areas dubbed "night clubs," where cats—those that survived over the two to three lean months when terns are not on Ascension—await them and consume many. Generations of terns suffering from cat predation may have led to the greater synchrony in the breeding schedule on this island and briefer use of the night clubs (Ashmole 1963).

While birds may abandon a nest or a roost site after having been disturbed by predators, few species may have any ability to detect danger in advance. Some species are more sensitive than others, and some avian senses trigger more of a response than others. Experiments show that vision leads to avoidance or alertness more than scent, which other experiments have demonstrated is indeed functional for many other purposes in birds. In winter experiments in Poland, Great Tits (*Parus major*) did not use nest boxes that contained a few dog or cat hairs or bits of tit feathers. Whether they rejected these boxes because they recognized the prior presence of a potential predator or were responding to an indication of potential parasites or pathogens could not be determined, but the hair and feathers would, in any case, have been left by a predator (Ekner and Tryjanowski 2008). Ring Doves (*Streptopelia risoria*) exposed to a tame ferret (*Mustela furo*) before sleeping opened their eyes more often subsequently while sleeping than ones not exposed to it. Doves that were in ones and twos when exposed to the ferret later opened their eyes more often than doves exposed in groups of up to six. All the exposed birds kept their eyes open longer each time they opened them than the birds that had not been exposed to the ferret (Lendrem 1984.)

Scent is another signal of risk, although it may be a lesser concern in certain situations. A study with Great Tits in captivity showed that they do avoid nest boxes with predator scent. Other studies with wild birds in the breeding season also demonstrated that tits did avoid predator chemical cues inside or near nest boxes. But experiments with Great Tits and Blue Tits (*Cyanistes caeruleus*) in Spain and the Netherlands that tested whether they perceived the odor of a mustelid in a nest box at the time they entered it to sleep in the nonbreeding season found that the birds used these boxes at the same rate as the ones scented with lemon oil or with nothing, indicating that, if they perceived the odor, either they did not associate it with any danger or with sufficient danger to avoid. It may be that under natural winter conditions tits prioritize a roost site that has insulating benefits over the risk that one previously visited by a predator will be visited again (Amo et al. 2018). When Great Tits were exposed to the scent of a mustelid while sleeping, they did not respond differently from the test birds exposed to rabbit scent or none at all (Amo et al. 2011).

Insects and Other Parasites

Sleeping birds attract many biting and parasitic insects. This may be another reason, beside thermal benefits, why so many birds sleep with their exposed skin concealed: the head tucked beneath scapular feathers and the feet raised so they are covered by feathers of the belly. Avoiding both insects that seek a drink of blood or a bite of flesh, and parasites that attach themselves for longer periods, are among the factors that at least some birds consider when choosing their roost site.

Tits and perhaps other cavity nesters are able to discern and choose among roost sites that may expose them to parasites. Great Tits in Switzerland, when offered a choice between a clean nest-box and one containing an old but parasite-free nest, chose them equally. When the choice was between a parasite-free nest box and one infested with blood-sucking hen fleas (*Ceratophyllus gallinae*), they chose the clean box. Finally, when there was only an infested box, most birds roosted outside (Christe et al. 1994). Tits that find hen flea infestations in their nests lose sleep dealing with them. Females monitored in parasite-free nests slept 73.5 percent of the time, while those in nests with parasites slept only 48.1 percent of the time. These spent 27.1 percent of their waking time devoted to nest sanitation; those in the clean nests only 8.3 percent (Christe et al. 1996).

Birds that roost in the open are vulnerable to other parasites. Palm Warblers (*Setophaga palmarum*) and Prairie Warblers (*S. discolor*) wintering in dry scrub vegetation in the Dominican Republic are exposed to the scaly leg mite (*Knemidokoptes jamaicensis*), which causes warty lesions on the legs, additional skin damage, and overgrowth of toenails that can affect a bird's ability to perch. Here, the warblers roost communally; this may enable mites to move from one bird to others, while the same warblers roosting solitarily in pines on other parts of the island do not carry the mites. During particularly dry winters, as many as 25 percent of Palm Warblers roosting in scrub are infested; over two successive winters, no infested warblers of either species returned the following year. Since other, year-round residents in the Dominican Republic are also infested, it is not likely that the warblers brought the mites with them (Latta 2003).

For other birds, however, communal roosting may reduce the likeli-hood of being parasitized, just as it reduces the chance of being predated. The mosquito *Culex pipiens*, which carries West Nile virus, is attracted to birds by their emissions of carbon dioxide, so large assemblages of birds roosting together become a magnet. Since the mosquito prefers to feed on species roosting in the canopy, House Sparrows (*Passer domesticus*), which often roost communally in shrubs, are only infrequently infected. At three communal roosts of American Robins in Washington, D.C., and nearby Maryland, mosquitoes were found to be most numerous higher in trees. While the number of mosquitoes present at these roosts was great, the actual density of mosquitoes per bird was several hundredfold fewer than for birds roosting alone (Janousek et al. 2014).

For some birds, avoidance of parasites at the roost is part of a web of other ecological factors, including the responses of neighboring bird species, exposure to other predators, and the population cycles of these predators' alternate prey. In the southwestern Yukon, from mid-June through August, Great Horned Owls shift their typical roost from concealed places in trees to exposed ones on the ground to avoid blood-sucking black flies (*Simuliidae*), which transmit avian malaria (*Leucocytozoon* spp.). These flies are then densest at mid-height in trees, where the greatest number of birds of all kinds are found. Black flies are scarce near the tops of trees, but owls avoid roosting there because they are harassed by shorebirds and gulls from nearby ponds as well as by various resident hawks and Common Ravens (*Corvus corax*). By roosting on the ground, they avoid harassment, but they are vulnerable to lynx (*Lynx canadensis*) and coyote (*Canis latrans*), espe-cially during the years when snowshoe hares (*Lepus americanus*) are at the low end of their ten-year population cycles (Rohner et al. 2000).

Birds are also of interest to moths. A few are known to insert their long proboscis into the eyes of birds to obtain nutrients, a feeding method much more frequently deployed by moths on reptiles and mammals. Moths feed on animal tears primarily to extract sodium and proteins. Lachryphagy on birds has been observed only a few times, likely because the chance of coming upon a sleeping bird in the dark and finding a moth on it is far lower than the phenomenon itself. It was first observed in

The rarely witnessed phenomenon of a moth feeding on tears from a sleeping bird, here a Black-chinned Antbird, was discovered by chance by a herpetologist doing a night survey in the Amazonian rainforest of Brazil. (With permission, from de Lima Moraes 2018.)

Madagascar, where in 2004 one moth species was found on a Common Newtonia (*Newtonia brunneicauda*) and a Madagascar Magpie-Robin (*Copsychus albospecularis*). The moths remained and continued feeding on each bird another thirty to thirty-five minutes after being discovered; neither bird showed any sign of disturbance (Hilgartner et al. 2007).

More recently, observations have come from the American tropics. In 2012, a Ringed Kingfisher (*Megaceryle torquata*) roosting on a branch at the edge of a river in Amazonian Colombia was found with a moth on its neck, the moth's proboscis inserted under the bird's nictitating membrane at the upper corner of its left eye (Sazima 2015). Finally, in 2017, two Black-chinned Antbirds (*Hypocnemoides melanopogon*) were found near midnight during a night survey in Amazonian Brazil, each with a moth of the same species on its neck. The moths' long proboscis enabled them to alight far from the host bird's eye, avoiding any disturbance or flight. One moth was seen moving its proboscis toward the eye, sometimes resting it inside the eye and feeding on the secretions. Each antbird remained immobile, seemingly undisturbed. Lachryphagy may decrease the fitness of host animals by increasing the chance of ocular diseases, but this is unknown (de Lima Moraes 2019).

Interspecific Competition

Thermal benefits, safety from predation, and capacious spaces for eggs and young are among the reasons certain roost sites and potential nest sites are highly sought-after, carefully guarded, and vigorously defended against conspecifics. Scarcity of ideal sites leads also to interspecific competition. Some birds systematically destroy the cavity roosts of smaller neighbors and drive away other birds roosting nearby. This may reduce the number of other birds that could draw the attention of predators.

Cavity Excavators

Woodpeckers, creators of what for themselves and later for many other birds are the most valued roosts in any wooded landscape, face competition themselves. Interspecies competition may not, however, require actual confrontation. In a Malaysian forest, a White-bellied Woodpecker (*Dryocopus javensis*) that had been roosting in the cavity of a stub twenty meters off the ground chose one night to abandon that tree when a Great Slaty Woodpecker (*Mulleripicus pulverulentus*) almost twice its size arrived shortly before dark and occupied an old woodpecker hole far higher in the stub. The next evening, the White-bellied Woodpecker did not return and the larger bird again had the tree to itself. In the same area, a White-bellied Woodpecker was seen enlarging and thereby destroying the roost hole of a Crimson-backed Woodpecker (*Picoides cathpharius*) less than half its size (Short 1973, p. 357, 288).

Many other woodpeckers are known for this habit. In eastern North America, Downy Woodpeckers (*P. pubescens*), as the smallest woodpecker in the woods, have to contend with other woodpeckers and cavity roosters, and they are sometimes aggressive in chasing them away, especially when the hole is being used as a nest. But both Hairy Woodpeckers (*Leuconotopicus villosus*) and Red-bellied Woodpeckers (*Melanerpes carolinus*) sometimes eliminate the competition by enlarging the Downy's entrance hole, even though they do not later use this cavity for themselves (Kilham 1983, p. 7).

Some barbets also excavate their own cavities for roosting and nesting. Like woodpeckers, they, too, enlarge and make unusable the nearby cavities of smaller barbets and do not then use the cavity themselves. In Africa, Anchieta's Barbet (*Stactolaema anchietae*) drives away two smaller barbet species when these are within forty meters of a roosting hole. It also attacks various orioles, shrikes, and swallows—none of which use tree holes—when these fly or perch near this barbet's cavity. The widespread Asian Coppersmith Barbet (*Megalaima haemacephala*) actively defends its hole against woodpeckers and drives parakeets away from the vicinity. Coppersmiths spend more than 51 percent of their time roosting, which suggests guarding against potential usurpers; they are nevertheless sometimes evicted by the larger Blue-throated Barbet (*M. asiatica*) (Short and Horne 2001, pp. 67, 145).

Crimson-throated Barbets (*M. rubricapilla*) in southern India and Sri Lanka were similarly observed over three months spending almost twelve and a half hours of the day at their roost cavity. In one tree, another cavity was occupied by a pair of White-cheeked Barbets (*M. viridis*), which drove them off when nesting season began. One of these White-cheeked Barbets was then pursued for three evenings by a Brown Hawk Owl (*Ninox scutulata*); the barbet escaped by flying into its roosting hole (Short and Horne 2001, pp. 284, 239).

Cavity Adopters

In the Neotropics many birds, from toucans and owls down to wrens, depend on woodpeckers to create cavities of appropriate dimensions that they later occupy. Some watch and wait until the woodpecker young have fledged, while others try to occupy the hole immediately, even before the woodpeckers have finished it. In Costa Rica, Fiery-billed Aracaris (*Pteroglossus frantzii*) monitor the progress in the nests of Lineated Woodpeckers (*Dryocopus lineatus*), one of the few woodpeckers that excavates cavities large enough to hold the groups of aracaris that habitually roost together. At one nest, aracaris came regularly to see if it was occupied; on the third night after the young woodpeckers had fledged, five aracaris moved in (Skutch 1958).

At another cavity used at night by three Fiery-billed Aracaris, a pair of Masked Tityras (*Tityra semifasciata*) began to bring material during the day to what they intended to use as their nest while the aracaris were nowhere near. For several days, the tityras added material, and each night, the aracaris continued to use the cavity, despite the tityras darting at them as they entered. When the cavity had become so filled that the aracaris could no longer fit comfortably, one removed some of the tityras' nest material before the three settled in. Eventually, the female tityra probably laid an egg in the cavity that was likely crushed by the aracaris, and within a few days, both sets of birds abandoned the hole. Two weeks later, however, the tityras occupied the site again and proceeded to nest without disturbance (Skutch 1958).

Green Woodhoopoes are part of a hierarchy competing for tree cavities in the East African savanna. They are dominated by Lilac-breasted Rollers (*Coracias caudata*), fight over cavities most often with Blue-eared Starlings (*Lamprotornis chalybaeus*), and are dominant over Grey Woodpeckers (*Dendropicos goertae*) and Hoopoes (*Upupa epops*) (Ligon and Ligon 1978).

In an experiment with Great Tits and Blue Tits in aviaries with some nest boxes that had large entrance holes and others with small holes, 76 percent of the Blue Tits occupied the larger boxes when no Great Tits were present. When a Great Tit was introduced to the aviary, half the Blue Tits switched to the boxes with smaller entrances, even if the Great Tit did not sleep in any of the boxes. In an aviary containing several Great Tits, none changed their roost site when a Blue Tit was introduced. During thirty hours of observation in eighteen experiments there were no aggressive interactions between the two species. In the wild, however, as early as September, long before Great Tits will use them for the winter, they defend large-holed boxes against Blue Tits. For a Blue Tit found in one of these boxes, the encounter is sometimes lethal (Kempenaers and Dhondt 1991).

Open Roost Sites

Protected or concealed sites other than cavities can also be contested. A female Spot-crowned Euphonia (*Euphonia imitans*) that Alexander Skutch had watched for two years roosting in a fold of hanging liverwort three meters over the ground was challenged one evening by a Bananaquit (*Coereba flaveola*) that had lost its dormitory. The Bananaquit used a few straws to narrow the entrance and then occupied the roost. It pecked at the euphonia. The two fought and tumbled twice to the ground, and the euphonia returned alone to its roost. Over the next few days, however, the Bananaquit added so much material to the entrance that the euphonia abandoned it (Skutch 1989b, p. 35).

In expansive roosts of several species, sometimes with similar or widely different ecologies, it may be difficult to determine the extent of actual competition versus subtle site preferences or long-established hierarchies. In Singapore, four species of starlings and mynas roost together in large groups of trees with dense foliage, but there is some segregation, as well as some conflict, among the four (Feare and Craig 1999, p. 152). In wetlands in Zimbabwe that host several species of finches and weavers, as well herons, kites, and doves, most of the smaller birds settle into the reedbeds, while the adjacent willows may not be sufficient to hold all the birds that prefer to roost in trees. Black-shouldered Kites (*Elanus caeruleus*) are dominant there over several dove species, but they give way to Hamerkops (*Scopus umbretta*), Southern White-crowned Shrikes (*Eurocephalus anguitimens*), and Gray Herons (*Ardea cinerea*) (Brooke 1965).

Even some seemingly limitless roost sites may lead to competition. Black Vultures (*Coragyps atratus*) and Turkey Vultures (*Cathartes aura*) have taken to roosting on electricity pylons, where they can perch closer to one another and gain some thermal benefits that do not come when they roost in trees. At a roost in south Texas, both species prefer the upper levels of the pylons, but in winter when these are crowded, Turkey Vultures usually give way to Black Vultures by moving to lower levels. The upper level is probably easier for vultures to land on and take off from, while birds on the lower level also suffer from a rain of droppings falling from above. Even though the birds on the lower level are not close enough to one another to gain any thermal benefits, they prefer to stay there rather than move to another pylon where they would be more isolated and perhaps at greater risk of predation (Buckley 1998). The same hierarchy plays out at pylons in Virginia; in spring and summer, however, when temperatures are warmer and trees provide more concealment, Turkey Vultures leave the towers when Black Vultures arrive, settling instead in nearby woods (Stewart 1978).

Elsewhere, birds that may be direct competitors or prey for others using the roost seem to share the space without incident. At a well-studied site in Michigan, wintering Northern Harriers and Short-eared Owls roosted in the same field, with some owls roosting among the harriers, arriving in the morning before the harriers had left and remaining in the evening after the harriers had returned (Craighead and Craighead 1956, p. 96). In Scotland, Willow Ptarmigan (*Lagopus lagopus*) and Black Grouse (*Tetrao tetrix*) sometimes use the same roost area as Northern Harriers through the winter, alarmed only when a harrier flies close over them. The roost seems not to be used for hunting by the harriers (Watson 1977, p. 249).

Non-avian Competitors

Insects and other animals also sometimes drive birds from their roost. Green Woodhoopoes leave their roost at dusk or at night at the approach of acacia rats. During the day, honeybees (*Apis mellifera*) sometimes take over these sites (Ligon and Ligon 1978). Honeybees can disrupt roosting habits even where they do not entirely drive birds away. In Lahore, Pakistan, a swarm of bees that settled under the eaves of a building with a colony of House Swifts (*Apus affinis*) prevented birds with nests closest to the swarm from reaching them. For the next three nights, the swifts came to roost earlier than usual, within several minutes of sunset. The following day, the bee swarm departed and the swifts resumed their normal roosting schedule, somewhat later (Razack and Naik 1965).

In southern Africa, Black Eagles (*Aquila verreauxii*) and chacma baboons (*Papio ursinus*) seek the same roost sites: shaded ledges on smooth rock outcrops. Troops of baboons often spend the night below the roosts or nests of eagles, on rock slopes with crevices. Sometimes, they climb up the rock to the ledge used by eagles, arriving there earlier than the birds. One pair of eagles was seen diving several times at the baboons. When the male eagle landed over the ledge, a baboon ran up to it, and the bird took off. After several additional fast flights by the ledge, the eagles abandoned it and moved to another rock 1.5 km away, using it regularly thereafter and making it their nest site two years later (Gargett 1990, p. 58).

Summary

The many physical and behavioral adaptations birds have evolved to reduce their vulnerability when asleep testify to the risks posed during these hours. Competition for the safest roosting sites may be a daily feature of a bird's existence, both for birds roosting alone as well as for those in flocks. Securing a preferred site before another bird occupies it may come with the cost of cutting short feeding opportunities. Where several species seek a limited resource, such as tree cavities, there may be battles among them, as well as competition, if not combat, with other animals, from colonial insects to primates.

The actual extent of nocturnal predation at individual or communal roosts is challenging to measure. Observations of systematic rather than incidental roost predation are most available for species that roost in large flocks. These magnets for predatory birds and other animals attract owls and falcons at dawn and dusk, various nocturnal mammals, and many raptors during the day. Birds and other predators adjust their own timetable when migrating or wintering populations of small birds are in their region, not bothering to patrol roosting sites during vacant hours or seasons. Similarly, they can follow the same tidal cycles as shorebirds that at high tides roost in the most vulnerable locations. At some roosts, the collective predation over a season may total as much as 20 percent of the roost's initial population. When selecting roost sites, flocks choose those with most shelter or most visibility, but birds roosting alone have less ability to anticipate a predator's attack. In experiments, tits given the choice between boxes with visual signs of a predator and those without will choose the boxer that seems safer, but they do not use the scent of a predator to make the same choice.

Sleeping birds attract many biting and parasitic insects. Many birds reduce their exposure by sleeping with the head tucked beneath scapular feathers and the feet raised so they are covered by feathers of the belly. Avoiding both insects that seek a drink of blood or a bite of flesh and parasites that attach themselves for longer periods is likely a reason many birds, especially cavity users, rotate among several sites and defend these against conspecifics and other birds. A few moths specialize in drinking the tears of sleeping birds to get sodium and proteins; whether this causes any harm to the birds is unknown.

Birds that use scarce roosting sites like cavities or other sheltered places must compete for these with conspecifics and with other birds seeking the same site. Woodpeckers sometimes destroy the cavities made by smaller neighbors, and other cavity users drive away competitors, even nesting occupants, from a desirable hole. This may reduce the number of other birds that could draw the attention of predators. Both for cavities and exposed perches there is often a hierarchy for dominance and possession among several competing species, while elsewhere, birds that may be direct competitors or prey for others using the roost seem to share the space without incident. Roosting birds also sometimes need to vacate a favored roost for swarms of social insects and for aggressive mammals.

10

Human Impacts

Birds have long taken to roosting in towns, cities, and industrial facilities as well as in agricultural landscapes such as tree plantations and sugarcane fields. All these may provide birds with greater warmth, shelter, or safety from predators than more natural environments. At the same time, birds living in proximity to people are exposed to light and noise that have changed their sleeping habits. In more remote places, birds have suffered from human disturbance at essential roosting sites and have sometimes lost those sites entirely. Historically and today, the birds that roost in large aggregations have been harvested for food, often at unsustainable levels, and the proliferation of invasive animal species all over the world has made some birds easy prey where they sleep. Conservation initiatives need to be aware of how birds spend the part of their day that they are least visible but likely most vulnerable.

Artificial Light and Noise

Since the most important factor governing the sleep schedule of birds is changes in light level, birds that live where artificial light extends the day at its beginning, end, or both are likely to show changes in behavior. They may take advantage of the unnaturally longer days to continue foraging, to give more time to singing and courtship, or to begin breeding earlier in the season. Some birds have responded to urban noise by

shifting their waking or singing times to quieter hours when, in more natural conditions, they might have been sleeping. Few studies, however, have yet shown that extra light or noise actually affect survival or productivity compared with birds in natural regimes. Laboratory experiments, however, have demonstrated that there can be deleterious effects on the quality of sleep and basic physiological processes.

Extended Feeding Hours with Extra Light

At all latitudes, the extension of sufficient visibility for feeding may affect birds living through the very short days of winter; migrants refueling as voraciously as they can; and equatorial residents extending their day beyond its usual strict twelve-hour limit. In Bergen, Norway, at 60°N, where in midwinter there are eighteen hours of darkness, street lights have expanded the feeding time of some species, while others have not taken advantage of these to extend their waking hours. Of twenty-four passerine species studied in a suburb with street lights, European Robins (*Erithacus rubecula*), Eurasian Blackbirds (*Turdus merula*), and Eurasian Wrens (*Troglodytes troglodytes*) began their days four to five hours before sunrise, unlike their neighbors in darker areas. Only some of the local Great Tits (*Parus major*) and Blue Tits (*Cyanistes caeruleus*) rose that early, while the other nineteen species did not begin their activity any sooner than usual, at morning twilight. No species used the artificial light to continue foraging later in the evening, suggesting that the natural amount of daylight without the artificial morning extension provided enough hours for birds to feed themselves. There are likely ecological correlates with the change in behavior of the few species that took advantage of the street lights. All the early risers feed chiefly on invertebrates found on the ground, while the birds that did not change their hours are seedeaters, omnivores, or specialists in arboreal insects, for which food may either be sufficiently abundant or found beyond the range of the street lamps, so individuals that happen to live near artificial light have no incentive to begin their day sooner (Byrkjedal et al. 2012).

At 48°N in Munich, Germany, where during winter the days are longer than in Bergen, but shorter in spring and summer, a three-year study of individually tracked Eurasian Blackbirds in forest, city parks, and the business district during February through June found that the two sets of birds in urban areas were most active before civil twilight only from mid-March to mid-April. But compared with the much shorter Norwegian midwinter days, here during spring blackbirds did not advance their schedule by more than an average of eighty-five minutes. This early part of the breeding season is when they are most engaged in singing and courting. During the other months, the days of the urban birds closely matched those of the forest birds; in those months, the urban birds may have had less reason to take advantage of the extra hours of light. Throughout the test period, only blackbirds in Munich's business district continued activity later at the end of the day than those in more natural environments (Dominoni et al. 2014).

Artificial light also attracts certain prey types such as flying insects. For the blackbirds feeding in winter on slow-moving arthropods living on the ground, often under leaf litter, this is less relevant than for foliage gleaners and specialists in aerial insects during warmer months. The intensity and extent of the illuminated area must also influence the response of birds to this novel feeding opportunity. During nights from August into October when the area around the football stadium of Cornell University in Ithaca, New York, is lit with 156 1500-watt halide bulbs, Eastern Phoebes (*Sayornis phoebe*), Gray Catbirds (*Dumetella carolinensis*), and large numbers of at least fifteen species of migrating warblers sometimes feed in the adjacent trees and on the ground, picking insects off foliage and sallying for flying insects, mainly moths, that are attracted to the lights, which on some nights are lit until 2 a.m. During the day, these birds are not present on the grounds; at night, this feeding opportunity may attract them from other nearby wooded areas (Lebbin et al. 2007).

In the tropics, some resident birds have been seen at night pursuing insects drawn to street lights. At Las Minas de San Cristobal, El Salvador, a Turquoise-browed Motmot (*Eumomota superciliosa*) came regularly to a garden to feed on insects attracted by lamps, arriving forty-two to fifty-two minutes after sunset and remaining as many as thirty-nine minutes (Thurber and Komar 2002). Similarly, in San José, Costa Rica, a Blue-crowned Motmot (*Momotus momota*) was observed catching several sphinx moths at 9:30 p.m. near street lights (Solano-Ugalde and Arcos-Torres 2008). Motmots are crepuscular feeders, but in Brazil, many hummingbirds and a few flycatchers have been found at various times between 8 p.m. and 3:30 a.m. feeding at flowers and pursuing insects in well-lit gardens and near street lights (Sick and Teixeira 1981).

Shorebirds, already sometimes nocturnal feeders, depending on tidal cycles, also take advantage of the lights along roads adjacent to estuaries to feed additional hours on both dark and moonlit nights. The birds may use the artificially extended foraging opportunities at the expense of hours they might be resting. At several places around the Tagus Estuary in Portugal, six species of wintering shorebirds were observed while foraging within the range of road lights. The visual feeders among these species significantly increased their prey intake, compared with conspecifics in an area without artificial light. The species that used a mix of visual and tactile feeding increased their extent of visual feeding and had somewhat higher rates of success. The average of the two sectors was an 83 percent improvement in prey intake rate. Potential negative effects of feeding at illuminated areas, however, include increased risk of predation and exposure to pollutants that may be at higher concentrations near urban or industrial sites (Santos et al. 2010).

Impacts of Light and Noise on Singing Hours

Since most birds sing early in the morning, often before light, or before light levels enable them to feed, it is not surprising that artificial light prompts them to advance or extend their singing hours. Some species that frequently sing at the end of the day or after nightfall then also sing more in the day's final hours. As with artificial light's influence on activity, the effects on song may be strongest where the differences between the extent of natural and artificial light are greatest. The winter study in Norway found that robins in illuminated territories sang more during the night than birds in dark territories. This may have been to repel other robins from approaching their territory, with its extended feeding hours; robins there were seen feeding and chasing other robins during the night (Byrkjedal et al. 2012).

A breeding season study comparing the onset of song in natural and artificially lit habitats for five species in Finland, Germany, and Spain found that birds began singing earlier at all latitudes where there was extra light, with the exception of Eurasian Blackbirds and European Robins at 65°N in Finland, where the days began earliest and there was little or no actual natural darkness. Great Tits, Blue Tits, and Chaffinches (*Fringilla coelebs*) began singing earlier everywhere. The effect of artificial light was strongest in the Great Tit, which began thirty-five minutes sooner, and weakest in the Chaffinch, only seven minutes sooner (Da Silva and Kempenaers 2017). A companion study in Germany confirmed that for all these species it was the artificial light rather than noise levels from traffic that led to earlier singing; on weekends, when there was no traffic on the roads adjacent the forests where the birds were recorded, they began singing at the same time as on weekdays (Da Silva et al. 2014).

In contrast, near Sheffield, England, European Robins in illuminated areas that were relatively quiet did not sing at night, while those in particularly noisy areas did sing into the night. At these noisy sites, some of which were very dark and others illuminated, the daytime noise levels were about ten decibels higher than at the other places where the robins were also recorded. It has not been tested whether these robins were singing at night to reduce the time spent singing against acoustical competition or taking advantage of quieter conditions to continue singing longer than they would otherwise (Fuller et al. 2007). Similarly, in Singapore, the communal roosts of two species of mynas settle down for the night at different hours, depending on the noise level around them. Mynas at a roost near a highway continue singing and calling for ninety minutes longer than mynas in trees on a suburban street (Nee and Yeo 1993).

More broadly, a study of ten European species living near airports in Germany and Spain found that, compared with the same species in quiet areas, during spring they all began singing well before sunrise, thereby avoiding competition from the start of flights at 6 a.m. The birds in Germany, where there was more daylight before airport operations began, did not advance their singing initiation as much as birds in Spain. The species, mostly seedeaters, that normally begin their singing latest in the morning and therefore closest to the start of airport traffic, advanced their singing more than the insectivores that routinely begin their day at lower light levels. Earlier singing by additional species created a greater acoustical overlap among species singing near airports than in quiet habitat. This overlap may reduce the effectiveness of singing while also enabling nocturnal predators to locate the singers. Whether the birds that sing earlier near airports are actually anticipating the noise from takeoffs and landings and therefore adjusting their behavior, or are simply individuals within each species with a tendency to sing earlier, remains to be proven (Gil et al. 2015).

Impacts of Light on Breeding and Health

The effects of additional light from artificial sources extend beyond daily activities; they also affect annual cycles. Artificial lights may prompt birds to initiate nesting earlier and may affect nesting success. A two-year breeding season study in the Netherlands that installed street lights in a wooded area and kept them running from sunset to sunrise found that, in the first year only, Great Tits with territories in the newly lit area laid eggs about five days earlier and produced smaller clutches than those in dark areas. In the second year, temperatures were much warmer and all birds laid earlier. The light treatment did not affect the probability of brood failure or the number of chicks fledged in year one, but in the second year, tits farther from lights fledged fewer young. The lights had no effect on any aspect of the breeding of the other species studied—the Pied Flycatcher (*Ficedula hypoleuca*), also a cavity nester—or on the year-to-year survival of either species (de Jong et al. 2015).

A seven-year study of Blue Tits in Austria found that females near street lights began laying eggs 1.5 days earlier than those far from artificial light, but they did not lay more eggs. The extra light may have accelerated gonadal development and formation of eggs. Males in these territories were twice as successful in obtaining extra-pair mates as males in forest areas, most significantly in first-year males, which is contrary to usual patterns. The success of the males in lit territories may be because they began singing earlier, an indication of male fitness usually found in older males. It is possible, therefore, that female choice of first-year males for extra-pair copulations is maladaptive (Kempenaers et al. 2010).

Research is just beginning on how artificial light may affect the reproduction and general health of birds. Some results show only modest impacts, others more significant ones. In experimental groups of Great

Tits in the Netherlands exposed to either white, red, and green lights or to no lights at night, the birds exposed to white lights were most active during the night and between March and May showed the greatest decreases in oxalic acid concentrations, an indicator of sleep deficiency. During March, but not May, the white light birds also had the highest rates of avian malaria, but what caused this could not be determined; the local mosquito vector for malaria is only attracted to lights in autumn. None of the light exposures, however, affected levels of daytime activity or success in fledging that year's subsequent broods (Ouyang et al. 2017).

Results of Experiments under Controlled Conditions

The limited number of studies of birds exposed to greater amounts of light and noise at night in their natural environment have thus far not demonstrated significant effects on survival or reproductive success. Australian Magpies (*Cracticus tibicen*) in various parts of Melbourne and nearby quieter areas living in sound environments ranging from 65 decibels A to 40 decibels A—equivalent, respectively, to a noisy restaurant and a library—were found to do equally well on a variety of cognitive tests; individuals approximately 290 days old did better than those about 160 days old (Connelly et al. 2022). This set of experiments did not state whether artificial light levels differed at the various sites but, if it did, these did not affect performance either. In contrast, Great Tits in Estonia exposed during the day to a mounted Sparrowhawk (*Accipiter nisus*) simultaneously with various loud urban noises at different frequencies were least able to generate attention and mobbing behavior from their neighbors when the experimental noise was in the same frequency range as the tits' alarm calls. In this daytime situation, the noise can affect fitness through increased predation on adults or their young (Tilgar et al. 2022).

Different kinds of artificial light may vary in their impacts. As the world moves to more energy-efficient forms of lighting, the effects should be considered. In New Zealand, the retrofitting of streetlights in Auckland from high-pressure sodium (HPS) to white light-emitting diode (LED) provided a natural experiment. LED lights prompted introduced

Common Mynas (*Acridotheres tristis*) to begin singing several minutes earlier, while native Tūī (*Prosthemadera novaeseelandiae*) started singing forty-six minutes later than when they lived near HPS lights. For the native Silvereye (*Zosterops lateralis*) the new lights did not affect the initiation of song in the morning (McNaughton et al. 2021).

Florida House Sparrows (*Passer domesticus*) experimentally exposed to different types of light for two weeks had different responses to West Nile virus. The birds exposed to broad-spectrum (3000 Kelvin) light suppressed production of melatonin, which, among other functions, coordinates and controls immune defenses. Exposure to broad-spectrum/blue-rich (3000–5000 Kelvin) light did not reduce viral resistance, but it did increase West Nile virus mortality rates; these birds died from the virus while having lower viral burdens than individuals in the control group that were not exposed to any artificial light at night. Amber-hue (1800 Kelvin) light, however, increased resistance to the virus by maintaining lower virus burdens for shorter periods of time. These results match those from studies of artificial-light impacts on a wide variety of other animals that all show that amber-hue light has the least effect on health or behavior (Kernbach et al. 2020).

Experiments with a variety of birds looking at the effects on sleep itself as well as on other physiological processes show that the impacts of the type of artificial light vary by species as well as by the hours and intensity of the light to which they are exposed. Domestic pigeons (*Columba livia*) and Australian Magpies exposed to light at urban levels slept less intensely and less overall, favoring slow-wave sleep over rapid eye movement sleep, and they had more fragmented sleep later when the lights were switched off. In pigeons, the effects were similar for white and amber light, while for the magpies the amber light did less to alter sleep patterns. In the day and night following their exposure to extra light, pigeons slept more, recovering some but not all the sleep time they had lost, while magpies did not have any extra sleep (Aulsebrook et al. 2020).

Laboratory experiments that can separate the potentially interconnected effects of light, noise, temperature, and additional food sources in urban settings may not directly answer questions about survival or reproductive success, but they have revealed other impacts on sleep and physiological processes that might in turn affect survival and productivity.

When exposed in captivity to levels of light comparable to urban situations, Eurasian Blackbirds taken from both urban and rural locations advanced the onset of singing by two hours; the night lights suppressed their melatonin concentrations, leading to earlier activity. Over the longer term, male blackbirds in the laboratory exposed to light at night had an earlier peak in testosterone levels, developed their gonads three weeks sooner, and began their post-breeding molt earlier than birds held in dark conditions. In natural situations, an acceleration of breeding readiness stimulated by lights may create a mismatch with environmental conditions necessary for success (Dominoni 2015).

In their second year under these conditions, the light-exposed blackbirds showed more deleterious effects: they did not raise testosterone levels, failed to develop their gonads, and had an interrupted molt. The physiological link for these phenomena is not known, but it is possible that the continual exposure to dim light at night eliminated the stimulus that natural seasonal variations in day length generate. Another potential impact that was not monitored is whether greater activity at night, resulting in less sleep, is compensated for during the day by intermittent sleep bouts; these may not have the restorative effects that come from longer sequences of night sleep (Dominoni 2015).

Looking for other physical impacts, exposing Zebra Finches (*Taeniopygia guttata*) to dim light comparable to street lights was found to alter their daily patterns of activity and rest. It also induced nocturnal feeding, caused body fattening and weight gain, and reduced melatonin levels. Lipids increased in the liver, as did nighttime levels of glucose, suggesting an impairment of metabolism (Batra et al. 2019). Further significant potential damage from reduced melatonin production has been found using blood from Ring Doves (*Streptopelia risoria*). Melatonin, which attains its greatest concentrations in birds' bodies during the night, has been linked experimentally to the facility of certain cells to ingest harmful foreign particles, bacteria, and dead or dying cells and to enhance metabolism. Experiments showed that melatonin together with corticosterone was more efficient than either hormone alone in stimulating immune responses (Rodriguez et al. 2001). Thus, a bird with low melatonin levels due to significant exposure to artificial light may have reduced metabolism and immune effectiveness, not only at night but also during the day when corticosterone is at high levels in the bloodstream.

Experiments such as these reveal some of the effects that exposure to light beyond birds' normal regimes may have on a wide range of physiological processes in addition to sleep itself. These experiments, however, may subject birds to more hours or greater intensity of light than any wild bird would experience, since in nature birds can reduce their nocturnal exposure to lights, either by moving away or by roosting in foliage or other situations that block a nearby source. Thus, the results of laboratory research showing significant physiological impacts from extra light (or noise, which has not been tested in similar ways) may not in fact be felt by many wild birds.

Roost Disturbance

Large and conspicuous roosts have attracted predators throughout evolutionary history, and humans have likely been among the predators from the time they devised means to catch birds. Other birds suffer roost disturbance because people pass too close to the roost and because the very material on which birds roost is sought. The invasive species that people deliberately or inadvertently introduce to regions where birds have no natural defenses have further contributed to the disturbance and loss of roost sites and sometimes to the decline and extinction of birds themselves.

Harvests of Roosting Birds

The harvests of vast numbers of roosting birds is today probably less significant than in the past, because laws and regulations now protect birds in many parts of the world, and also because some of the most desirable species no longer exist in such large aggregations. Some no longer exist at all. In North America, Passenger Pigeons (*Ectopistes migratorius*) were hunted at all hours and all seasons, with nets, traps, and guns during the day, while at night fires were set under their roosts, sometimes using sulfur to suffocate the birds, a practice brought from Europe, where it was used on roosting Ring-necked Pheasants (*Phasianus colchicus*). Other people simply burned the vegetation on the ground while knocking the birds off their perch with poles. When Passenger Pigeons roosted low enough, they could be taken by bare hands (Schorger 1955, pp. 167–168).

In some parts of the world, wild birds are still sought for food, and large roosts of palatable species, however small, continue to suffer. Ornithologists and others witnessing and documenting these activities are few, so accurate accounts are limited. Richard P. ffrench (1967) studying Dickcissels (*Spiza americana*) wintering in Trinidad during 1959–66 noted few instances at roosts of trapping birds for pets or shooting them to eat. In Venezuela, however, people eat wintering Dickcissels throughout their range. Children use slingshots and throw stalks of sugarcane at dense flocks arriving at their roosts. People enter the roosts at night to club the birds with sticks and bats; some drive vehicles across the path of low-flying flocks (Basili and Temple 1999).

The large Barn Swallow (*Hirundo rustica*) roost in Nigeria, where the natural predators include four species of hawks, three falcons, three owls, and a coucal, is also a resource for local people. Estimates of the human take over the seven months of each year that wintering swallows use the roost range from 200,000 to 462,000 per year. During one period of February through April, approximately 105,000 were killed by people, while over an entire wintering season approximately 2,500 were taken by avian predators (Bijlsma and van den Brink 2005). The recently discovered roosts of millions of Amur Falcons (*Falco amurensis*) in Nagaland, northeastern India, have been exploited by nearby communities. Before conservation measures were implemented, fishermen at a reservoir would string nets among the roost trees and kill an estimated 140,000 in less than two weeks at the peak of migration. Indian and international conservation organizations have worked with the local communities and today the falcons are protected, the first tourist facilities have been created, and additional major roosts have been found in the neighboring states of Assam and Manipur (Weidensaul 2021, pp. 316, 339).

Other birds are destroyed at roosts because they are considered pests. In Venezuela, some farmers in the grassland region where Dickcissels winter poison birds using chemicals, such as an aerial application of parathion, early in the morning, while the birds are still at their roost. One farmer estimated that this had killed more than one million Dickcissels on his property over the years. During March and April, flocks of migrating Barn Swallows and Bank Swallows (*Riparia riparia*) that also roost in the cane fields and might be beneficial to farmers die from the same poisonings. The 40 percent decline of Dickcissel populations in North America between the late 1960s and the 1990s may be correlated with the simultaneous increase of lethal controls during the same period in Venezuela as land conversion and crop production expanded. Since some roosting aggregations may contain 30 percent of the entire Dickcissel population, this species is especially vulnerable to severe losses from single incidents (Basili and Temple 1999).

Understanding resting and roosting habits can significantly reduce unintended mortality of birds. Millions of seabirds in all oceans are killed each year by drowning after being caught in gillnets or swallowing baited hooks set by industrial fishing fleets. The Japanese tuna fleet operating in the waters off of Australia uses fishing lines that may be 100 km long and hold 2,500–3,500 hooks. These are meant to sink beneath the reach of albatrosses and petrels, but many birds get caught when the lines are set and are still near the surface. During winter, however, seabird mortality can be reduced by 99 percent if the lines are set at night. In summer, when the nights are shorter, albatrosses and petrels are more active, especially when the moon is bright or full. Today, night setting of fishing lines is mandatory in Australian and New Zealand fishing zones and in those covered by the Convention of Antarctic Marine Living Resources (Brothers et al. 1999), but not in some other major ocean fisheries.

Disturbance and Destruction of Roosting Sites

As humans convert more of the Earth's surface to land uses serving their own purposes, the habitats and ecosystems birds need for all their life cycle are degraded or lost. For some species, however, roost sites, more than other features of the landscape they need, are particularly vulnerable to disturbance and destruction. Such sites may be critical year-round or only seasonally, when birds shift their roosting habits or their location to regions where they are in closer proximity to people.

SHOREBIRDS: BOAT TRAFFIC AND BEACHES

Roosting shorebirds are among the species most sensitive to disturbance, and those for which roost disturbance factors have been best studied. Beachgoers and their dogs and vehicles, as well as passing boats, can disrupt the rest these birds require between their intensive feeding efforts running through much of the day and the night. In many places, shorebirds have few sites available during high tides. On their northward migration in spring, the eastern population of the Hudsonian Whimbrel (*Numenius hudsonicus*) concentrates at a few staging areas on the Atlantic coast before making a long flight to its Arctic breeding range. In 2019 and 2020, at least 19,485 whimbrels were found roosting at night at the Deveaux Bank in South Carolina; this represents approximately 49 percent of the estimated eastern population and 24 percent of the entire North American population (Sanders et al. 2021). The feeding and resting conditions whimbrels find here make it unlikely they could shift to other sites if this bank were disturbed.

For migrating shorebirds that have less highly concentrated staging and roosting habits, finding an undisturbed place to roost on the Atlantic coast can still be challenging during summer, when most beaches are crowded with people. At Plymouth Beach, Massachusetts, on a five-kilometer sandspit heavily used by southbound shorebirds in July and August, the species that favor the beachfront rather than the mudflats are the ones most sensitive to people and to vehicles. During peak summer days, a thousand or more vehicles may be parked or moving along the beach. Each shorebird species has a different threshold to disturbance when roosting: of the three commonest, most Short-billed Dowitchers (*Limnodromus griseus*) leave the beach if ten to forty vehicles are nearby, for Semipalmated Sandpipers (*Calidris pusilla*) and Sanderlings (*C. alba*), one hundred vehicles. Even when fewer than twenty vehicles

For shorebirds migrating or wintering on beaches, like these Sanderlings, walkers and their dogs are often the too-frequent disrupters of their rest.

are parked on the beach, however, only 50 percent of the Sanderlings and Semipalmated Sandpipers that roost there when it is empty remain; if more than a hundred vehicles are on the beach, only 11 percent remain. At this site, closing the tip of the sandspit to beachgoers and their vehicles was recommended to ensure sufficient space for roosting shorebirds (Pfister et al. 1992). More than thirty years later, that recommendation has not been acted on, but designation of the Piping Plover (*Charadrius melodus*) as an endangered species has resulted in summer closing of its nesting area on Plymouth Beach to vehicles. This has reduced disturbance at high-tide roosting sites of the migrant shorebirds, but it has not prevented walkers and their dogs from accessing the area, while recent decades have seen a proliferation of boaters landing on the beach. Farther south on the coasts of the United States, both migrant and wintering populations of shorebirds that use beaches likewise encounter resident and vacationing beachgoers throughout the months beaches are in high demand.

Boat traffic is also a factor for the shorebirds that winter in tidal wetlands of saltmarshes and barrier islands and for those near harbors. At
the Cape Romain National Wildlife Refuge north of Charleston, South
Carolina, Red Knots (*C. canutus*) avoid roosts that have high boat traffic
within 1000 m, while Short-billed Dowitchers and Long-billed Dowitchers (*L. scolopaceus*) and Hudsonian Whimbrels avoided roosts where
boats pass within 100 m (Peters and Otis 2007). On the northeast coast
of England, just north of the Tees estuary, a special effort was made in
1991 to provide a high-tide roost site for wintering Purple Sandpipers
(*C. maritima*), Red Knots, and Ruddy Turnstones (*Arenaria interpres*)
when the stone pier the birds had used was replaced in a harbor by a new
pier and an artificial island specifically intended as an alternate roost
site. The increase in boat traffic to the new pier, however, led to a decline
in use by all three species, as much as 50 percent for the turnstones and
80 percent by knots (Burton et al. 1996).

More challenging to determine is to what extent these degrees of
disturbance are actually costly to resting shorebirds. On the Dee estuary,
between North Wales and Liverpool, several species shifted their high-
tide roost site to an isolated beach twenty kilometers away when the number of walkers, horse riders, and dogs disturbed them too frequently. They
returned to the Dee to feed, sometimes making the trip twice daily. For
Red Knots, each roundtrip from the distant undisturbed site required an
estimated 14 percent increase in daily energy expenditure (Mitchell et al.
1988). A study of shorebirds wintering on the northern coast of Australia
near Darwin found that Red Knots and Great Knots (*Calidris tenuirostris*)
that made ten alarm flights from their high-tide roost each day likely increased their energy expenditure 4.5–4.7 percent. For Greater Sand Plovers (*Charadrius leschenaultii*) and Lesser Sand Plovers (*C. mongolus*) the
increase was 7.5–7.8 percent. As elsewhere, knots were more sensitive to
disturbances, flying from their roost when a person was approximately
75 m away, while the sand plovers flew when a person was about 45 m
away. Several larger shorebirds, including two curlew species and two
godwits, routinely left the roost when people were more than 100 m
away (Lilleyman et al. 2016). At another roost site in northwest Australia
where the two knot species were disturbed more than three times per

hour by a mix of natural predators and humans, the combined costs of flying to roosts and flights due to disturbance there was estimated at 17.3–28.7 percent of their total energy budget (Rogers et al. 2006b).

At some sites, wintering shorebirds gradually become acclimated to human disturbance, flying only when approached more closely and, over time, settling more rapidly afterward. Other species such as nonmigratory Black Oystercatchers (*Haematopus bachmani*) use traditional roost sites on the Pacific coast of North America throughout a breeding season or for years. Repeated disturbance there may cause abandonment for months or years (Andres and Falxa 1995). The real test of whether the additional unnatural disturbance actually affects shorebird survival or fitness has not been measured. Most studies, however, include recommendations that the sites birds depend on, especially during the highest tides when alternatives are fewest, be protected by limiting human access.

LANDBIRDS AND DEMANDS FOR WOOD

In economically advanced countries, disturbance specific to roost sites— as distinguished from habitat loss—may come mainly from people pursuing leisure or vacation activities. Elsewhere, there is competition between birds and people for increasingly scarce resources. Even birds that roost inconspicuously or in small numbers are vulnerable at their roosts when these are resources valued for other purposes. A study of Green Woodhoopoes (*Phoeniculus purpureus*) near Lake Naivasha, central Kenya, was interrupted when people chopped down several large old trees in two territories to make charcoal. This eliminated the roost trees used by the woodhoopoes, which live in families that sleep together in cavities. One family group vanished within a few months, another flew two kilometers twice each day across grassland to a roost tree beyond its normal territory; it also eventually disappeared. No new groups colonized the areas that now lacked roost trees (Ligon and Ligon 1978).

Black Storks (*Ciconia nigra*) wintering in West Africa prefer to roost in tall dead or dying trees above surrounding vegetation. They rotate among roost sites approximately every four days, staying within four kilometers of their current feeding area. Both bush fires and tree cutting for firewood reduce their available roost sites, while clearing of land for agriculture eliminates the area available where new trees might grow to replace the ones removed (Chevallier et al. 2010). This situation is likely to be the same in much of the world where large trees are scarce and sought-after for firewood or construction.

Among small landbirds, individual tracking using transmitters has shown how dependent some species are on distinctive roost-site features in what otherwise seems suitable habitat with an abundance of potential roosts. As much as 36 percent of male Cerulean Warblers (*Setophaga cerulea*) in northern Alabama routinely roost beyond their home range of 6.7 hectares, and only 13 percent of males roosted in the core area of their territory. Since all the birds in this study occupied a heavily forested, undisturbed landscape, the roosting requirements must be very specific. At the same time, only one male of the ten that were tracked roosted in the same tree more than once. Trees near clearings and edges were not used, indicating that, for this sensitive and declining species, habitat disturbance, even of areas peripheral to their core territory, can reduce occupation (Carpenter and Wang 2018).

Invasive Species

After habitat loss, invasive species are considered the greatest threat to natural ecosystems and all the species in them, both plant and animal. Some invasives have a particularly direct impact on roosting birds. Around the world, introduced snakes, mongooses, mustelids, foxes, rodents, and cats attack birds when they are asleep. The impacts are global, but they are particularly significant on islands where birds had no predators until recent introductions. Few birds that evolved in places without predators, or none that attack at night, have been able to adapt.

CATS

On a global basis, of all invasive species, domestic cats (*Felis catus*), both feral and those let out of the house, are probably the greatest killers of roosting birds. While cats hunt at all hours, their nocturnal habits suggest that a large number of birds, especially those roosting or nesting on the ground or in accessible vegetation, are killed at night. In the continental United States alone, there are an estimated 80 million house cats and another 30–80 million feral cats. A variety of studies make them responsible for killing 1.3–4 billion birds in the United States each year. About 69 percent of these birds are killed by unowned cats. (Loss et al. 2013). The most recent annual estimate for Australia is 377 million birds killed each year (Woinarski et al. 2017), for China it is 2.6–5.5 billion (Li et al. 2021).

In contrast with the broad impact of cats on large landmasses, on islands with few bird species at high densities, the effects of nocturnal predation are magnified in scale and concentrated in impact. In 1948–49 five house cats were brought to Marion Island in the southern Indian Ocean to control the house mice (*Mus musculus*). By 1977, the cats had multiplied to 3,405 and were preying primarily on adults, chicks, and eggs of eight nesting seabird species. Cats took an estimated 455,000 birds each year, while mice were only 16 percent of their diet. By 1991, the cats had been eliminated (Carboneras 1992), but mice remain and have begun attacking chicks of Wandering Albatrosses (*Diomedea exulans*). A mouse eradication plan is scheduled to begin in 2025. On Ascension Island in the mid-Atlantic, a project from 2002–2006 cleared the last of the cats, descended from those introduced soon after 1815. As a result, seabirds, which once numbered tens of millions of several species, are reoccupying parts of the island where they used to nest and roost (Pitches 2013).

RATS AND MICE

Rats on islands prey on seabirds of all sizes. Roosting birds can respond to predators by moving—if any place on the island still lacks rats. Birds incubating eggs or brooding their young at night are always stuck in place. At night, albatrosses and other seabirds on some Pacific islands are attacked by Polynesian rats (*Rattus exulans*), which climb onto their back and bite into their flesh, creating deep wounds as they gnaw. The process and the seeming helplessness of the victims is well documented at Kure Atoll in the Hawaiian Leeward Islands. Laysan Albatrosses (*Diomedia immutabilis*) nest in open sandy areas, with the parents alternating incubation shifts and never leaving the egg untended. As many as twenty rats have been seen feeding simultaneously on one bird, with the albatross occasionally turning its head backward, but not otherwise moving; the rats are usually beyond its reach. Birds thus attacked are usually dead by the next morning. Several of the other thirteen seabird species nesting on Kure have also been prey for rats (Kepler 1967). The rats were finally eradicated in 1995 by the State of Hawaii's Department of Lands and Natural Resources.

On islands such as Gough Island in the South Atlantic that never had cats or rats, mice have also become significant predators of roosting birds. Eight million seabirds of twenty-four species nest on Gough. Two highly endangered species, Tristan Albatross (*D. dabbenena*) and Atlantic Petrel (*Pterodroma incerta*), are attacked at night by mice, with as many as ten feeding on a single live albatross chick, which is three hundred times heavier but unable to defend itself. Both these species have suffered extreme declines due to reproductive failure because mice kill so many of the young. Other more abundant nesters are also suffering (Wanless et al. 2007). The Royal Society for the Protection of Birds began a mouse eradication project in 2021; it has substantially reduced the mouse population and will continue until the last ones are eliminated.

SNAKES AND OTHER MAMMALS

Invasive nocturnal snakes and predatory mammals of many additional kinds attack birds while they sleep. Some have had complex behavioral effects on roosting birds. Perhaps it is another example of the intelligence of corvids that the critically endangered Mariana Crow (*Corvus kubaryi*) has taken to roosting in flocks since the arrival of brown tree snakes (*Boiga irregularis*) on Guam in the 1940s. The extra vigilance that comes from communal roosting, however, has not prevented the snakes from climbing trees and killing crows (Wiles 1998).

In Tasmania, the critically endangered Swift Parrot (*Lathamus discolor*) suffers heavy predation from the nonnative sugar glider (*Petaurus breviceps*), a small nocturnal possum introduced to Tasmania from Australia in 1835 that since the 1940s has spread across most of the main island. It enters tree cavities and kills nesting females and their nestlings while they sleep. This has led not only to a rapidly declining population but also to a strongly male-biased sex ratio, with intense competition among males for females. At sites with heavy predation, female parrots are likely to mate with more than one male. These nests have lower fledging success, on average one fewer chick than in nests with lower mixed paternity, perhaps because the males with which the female has copulated interfere with or disrupt the efforts the actual mate makes in feeding and caring for the young. Today, conservationists are working to control sugar gliders and to provide safer nest sites for the parrots (Heinsohn et al. 2019).

INSECTS AND VIRUSES

The Hawaiian honeycreepers illustrate how a configuration of invasive species synergizes to do damage to sleeping birds where individually they would not. Of the thirty-one species known from the time the islands were colonized from the United States in the nineteenth century, seventeen survive, with several of these endangered, some on the brink of extinction, possibly within a decade or less. Most have been victims of infections carried by mosquitoes, notably *Culex quinquefasciatus*, which seeks blood from birds mainly at night. It was first introduced to the islands in 1826 and then spread through the archipelago. Avian pox, *Poxvirus avium*, transmitted both by mosquitoes and direct contact, had already arrived sometime after the late 1700s. Avian malaria (*Plasmodium relictum*) came to O'ahu around 1865, borne by introduced Nutmeg Mannikins (*Lonchura punctulata*) and Spotted Doves (*Streptopelia chinensis*) from Asia. As these two species spread to other islands, they brought the pathogens with them, enabling these to be transmitted by the invasive mosquitoes to native birds, which had no natural immunity. The honeycreepers living at lowest elevations were the first to be lost; as the mosquitoes reached higher elevations—600 m by the 1950s, 1500 m by the 1970s—additional species were extirpated. Now, climate change is warming the islands and has enabled mosquitoes to continue their ascent, exposing more populations to avian malaria and likely rapid extinction. The honeycreepers were particularly vulnerable to mosquitoes because they had lost the habit likely present in their mainland ancestors of sleeping with head and legs covered by feathers, which reduces the area of exposed skin available to mosquitoes. Only a few of the surviving species have begun to evolve some resistance to both avian malaria and avian pox (Pratt 2005, pp. 166–168).

Management

In addition to inadvertent disturbance of roost sites, there are sometimes direct conflicts between people and birds where they roost. When birds are considered an agricultural pest, the challenge for farmers is how to prevent birds from consuming their crop while not damaging the crop itself. Where Dickcissels roost in Venezuelan fields of sugarcane, the farmers do not have the option of destroying the roost site. Elsewhere, various nonlethal methods may be used, such as making the roost unattractive to birds. When people in Iquitos, Peru, thought a quarter-million Southern Martins (*Progne modesta*) roosting in the trees of the Plaza de Armas were an ill omen for the future of the city, the mayor had the trees pruned of all their branches and the birds shifted their roost to somewhere beyond the city (Oren 1980).

It is likely that the Iquiteños were also disturbed by the noise, mess, and smell made by the martins at the plaza, a problem created by various flock-roosting birds in cities all over the world. At an undisturbed British winter roost of European Starlings (*Sturnus vulgaris*), the accumulation of droppings may exceed 30 cm over the season; in cities and towns, this must be removed frequently to keep sidewalks clean. Where starlings roost in trees, their droppings may scorch leaves and twigs and sometimes kill trees; the mere weight of birds on branches can cause them to break off. Large flocks roosting in commercial tree plantations may cause the death of 5 percent of the trees if the site is used for a full year; over longer periods, the mortality can reach 100 percent. In addition, the seeds defecated by starlings, and in North American roosts likely by blackbirds as well, will germinate and interfere with the commercial crop (Brough 1969).

The droppings at large roosts can have a direct impact on human health. Bird droppings carry more than sixty diseases, some specific to certain species or families of birds, others widespread. Most significant of these, bird droppings are a growth medium for the fungus *Histoplasma capsulatum*, which causes histoplasmosis, a skin infection, in humans. More rarely, it infects the lungs and can move elsewhere through the bloodstream. In southeastern Missouri near some mixed-species roosts of starlings and blackbirds covering twenty acres and holding eight to ten million birds, children were found to have substantially higher skin sensitivity to histoplasmin than children in nearby cities with no roosts. Among the children tested at a school that was only a few blocks from a roost, there was no difference between the children who lived on farms, perhaps already exposed, and those who lived far from farms, for whom the hours at school were their only source of exposure. Spores might have reached the school area on the wind, and the movement of millions of birds may itself have created air currents (Tosh et al. 1970).

Short of destroying the roost site or using lethal methods on the birds, there are ways to drive large flocks of birds from traditional roosts. In British experiments using various means to disperse starlings from their roosts, smoke and pyrotechnics were found to have limited impact. Broadcasting starling distress calls was more effective. This emptied roosts when played over one to three successive evenings as the birds arrived, but many birds simply settled a short flight away and then were reluctant to fly again even if distress calls were played nearby. Some original roosts were resettled within two weeks, either by the same birds or others, but more roosts remained unused the rest of the season. These may be reoccupied in subsequent years unless given the same broadcast treatment (Brough 1969). Broadcasts of distress and alarm calls of the relevant species have also been used with some effect on gull roosts near airports and on corvids (Bremond et al. 1968).

Mounted dead birds can also repel others from using a traditional roost. Experiments in Lancaster, Pennsylvania, used fresh crow carcasses, taxidermic mounts, and artificial crow models, all suspended conspicuously from trees, hanging upside down as though dead, to serve as visual signals to arriving American Crows (*Corvus brachyrhynchos*) and Fish Crows (*C. ossifragus*). The first birds to reach the roost increased their calling, which attracted others to the effigy. These birds flew off, succeeded by waves of newcomers, which each had the same reaction. At a roost of more than one hectare that usually held more than ten thousand crows, the birds eventually settled in trees far from the effigy. After dark, some returned to the original site. Smaller roosts needed only one effigy, larger ones several. All three types of effigy worked equally well, but at large roosts, added deployment of pyrotechnics and distress calls worked best to move roost sites away from the city (Avery et al. 2008).

On the positive side, some habitats can be modified to attract roosting flocks, either to draw them away from undesirable locations or to provide alternatives where natural sites are scarce. Research by the US Department of Agriculture on optimal features of mixed-species blackbird roosts concluded that, of all plants used as roosts, dense stands of tall bamboo (*Phyllostachys*) best provide the thermal benefits, protection from predators, and abundant perches necessary to attract large flocks. In the southern states, where there are many roosts of millions of blackbirds, a one-hectare stand of bamboo could support more than a million birds. In addition, bamboo maintains the necessary attributes for its entire lifespan, while trees grow out of the density blackbirds prefer. At some sites where blackbirds have adopted bamboo stands, they have returned year after year. The challenge is to draw flocks to these sites and away from places where the roosts are not wanted (Glahn et al. 1994).

In many places, shorebirds can similarly be provided alternative roost sites if their traditional ones cannot be protected from disturbance or are otherwise altered by development. Climate change adds urgency, since sea level rise and higher tides are already reducing or submerging some essential roosting areas. When the harbor near the Tees estuary replaced its pier and expanded its facilities, the artificial island created for the shorebirds met the needs of the Purple Sandpipers and turnstones, because these regularly roost on sloping rocks. Knots, however, need a flatter surface than was provided, so they can mass tightly as well as spot any approaching predators (Burton et al. 1996). In Australia and South Korea, mesh bags filled with empty oyster shells attached to float foams and roped together in clusters up to 25 meters wide are serving as artificial roost sites used by migrating and wintering shorebirds in wetlands where high tides cover all nearby natural sites. This has been especially valuable where recent development and conversion of wetlands have eliminated the higher ground that was previously available to shorebirds (Runwal 2020).

In the Tagus estuary, Portugal, wintering Dunlins (*Calidris alpina*) do not usually use feeding sites more than five kilometers from their high-tide roost. This limits the number that can exploit the intertidal flats on the estuary's northwest side. Scientists found that creating a roost in an old drained wetland area would solve the problem. It would reduce by 25 percent the average distance that Dunlins need to fly to potential feeding grounds and would substantially increase the use of nearby mudflats. Similar analyses of the distances that various shorebird species can easily travel between their feeding areas and most important roosts—those available during highest tides—can be used to locate sites that could be protected or landscaped to provide new roosts where existing ones are too distant or too small (Dias et al. 2006).

Summary

Sleeping birds, no more than waking birds, have not escaped the influence of expanding human populations, settlement, and land use. Birds living in proximity to people are exposed to light and noise that has changed their very sleeping habits. In more remote places, birds have suffered from human disturbance at essential roosting sites and sometimes lost those sites entirely. Historically and today, the birds that roost in large aggregations have been harvested for food, often at unsustainable levels, and the proliferation of invasive animal species all over the world has made some birds easy prey where they sleep. Conservation initiatives need to be aware of how birds spend the part of their day that they are least visible but likely most vulnerable. Success stories in controlling and eradicating invasive species like cats and rodents on islands, and management practices that can remove roosts from undesirable places, show that at least some of the human impacts can be resolved.

Light is the most important factor governing the sleep schedule of birds, so those living where artificial light extends the day at its beginning, end, or both are likely to show changes in behavior. The effects may be strongest where the differences between the extent of natural and artificial light are greatest—at high latitudes in winter. Some birds take advantage of the unnaturally longer days to continue foraging, to give more time to singing and courtship, or to begin breeding earlier in the season. Birds have responded to urban noise by shifting their waking or singing times to quieter hours when, in more natural conditions, they might have been sleeping. Few studies, however, have yet shown that extra light or noise actually affect survival or productivity compared with birds in natural regimes. Laboratory experiments, however, have demonstrated that there can be deleterious effects on the quality of sleep and basic physiological processes.

Large and conspicuous roosts have attracted predators throughout evolutionary history, and humans have likely been among the predators from the time they devised means to catch birds. Today, in regions where wild birds are a source of food, harvest at communal roosts can take an unsustainable share of the population. Other birds have suffered roost disturbance because people, their vehicles, and their pets pass too close to the roosts and because the land or the very material on which birds roost is sought. The invasive species that people deliberately or inadvertently introduce to regions where birds have no natural defenses have further contributed to the disturbance and loss of roost sites and sometimes to the decline and extinction of birds themselves.

Roost sites sometimes cause direct conflicts between people and birds. When birds are considered an agricultural pest, the challenge for farmers is how to prevent birds from consuming their crop while not damaging the crop itself. Elsewhere, large roosts create health risks. Short of destroying the roost site or using lethal methods on the birds, there are ways to drive flocks of birds from traditional roosts to more neutral locales.

BIBLIOGRAPHY

Allard, H. A. 1930. The first morning song of some birds of Washington, D.C.: Its relation to light. *Am. Nat.* 64(694):436–469.

Alonso, J. A., J. C. Alonso, and J. P. Veiga. 1985. The influence of moonlight on the timing of roosting flights in Common Cranes *Grus grus. Ornis. Scand.* 16(4):314–318.

Alonso, J. C., I. Abril-Colón, and C. Palacín. 2021. Moonlight triggers nocturnal display in a diurnal bird. *Animal Behaviour* 171:87–98.

Alonso, J. C., L. M. Bautista, and J. A. Alonso. 2004. Family-based territoriality vs. flocking in wintering common cranes *Grus grus. J. Avian Biol.* 35:434–444.

Amlaner, C. J., and N. J. Ball. 1983. A synthesis of sleep in wild birds. *Behaviour* 87(1–2): 85–118.

Amo, L., S. P. Caro, and M. E. Visser. 2011. Sleeping birds do not respond to predator odour. *PLoS One* 6(11): e27576.

Amo, L., G. Tomás, I. Saavedra, and M. E. Visser. 2018. Wild great and blue tits do not avoid chemical cues of predators when selecting cavities for roosting. *PLoS One* 13(9): e0203269.

Ancel, A., C. Gilbert, N. Poulin, M. Beaulieu, and B. Thierry. 2015. New insights into the huddling dynamic of emperor penguins. *Animal Behaviour* 110:91–98.

Andres, B. A., and G. A. Falxa. 1995. Black Oystercatcher (*Haematopus bachmani*). In *The Birds of North America*, no. 155 (A. Poole and F. Gill, eds.) Philadelphia: Academy of Natural Sciences; Washington, D.C.: American Ornithologists' Union.

Armstrong, E. A. 1954. The behaviour of birds in continuous daylight. *Ibis* 96(1):1–30.

———. 1955. *The Wren*. London: Collins.

Ashkenazie, S., and U. N. Safriel. 1979. Time-energy budget of the Semipalmated Sandpiper *Calidris pusilla* at Barrow, Alaska. *Ecology* 60(4):783–799.

Ashmole, N. P. 1963. The behaviour of the Wideawake or Sooty Tern *Sterna fuscata* on Ascension Island. *Ibis* 103b:297–364.

Atkinson, E. 1993. Winter territories and night roosts of Northern Shrikes in Idaho. *Condor* 95:515–527.

Aulsebrook, A. E., F. Connelly, R. D. Johnsson, T. M. Jones, R. A. Mulder, M. L. Hall, A. L. Vyssotski, and J. A. Lesku. 2020. White and amber light at night disrupt sleep physiology in birds. *Current Biology* 30:3657–3663.

Aulsebrook, A. E., T. M. Jones, N. C. Rattenborg, T. C. Roth II, and J. A. Lesku. 2016. Sleep ecophysiology: Integrating neuroscience and ecology. *Trends in Ecology and Evolution* 31(8): 590–599.

Avery, M. L., E. A. Tillman, and J. S. Humphrey. 2008. Effigies for dispersing urban crow roosts. In *Proc. 23rd Vertebrate Pest Conf.* (R. M. Timm and M. B. Madon, eds.), pp. 84–87. Davis: U. of California Press.

Back, G. R., M. R. Barrington, and J. K. McAdoo. 1987. Sage Grouse use of snow burrows in northeastern Nevada. *Wilson Bull.* 99(3):488–490.

Bagg, A. M. 1943. Snow Buntings burrowing into snowdrifts. *Auk* 60:445.

Ball, J. R., K. C. Lukianchuk, and E. R. Bayne. 2011. Nocturnal provisioning by Swainson's Thrush. *Wilson J. Orn.* 123(3):508–514.

Ballerini, M., N. Cabibbo, R. Candelier, A. Cavagna, E. Cisbani, I. Giardina, A. Orlandi, G. Parisi, A. Procaccini, M. Viale, and V. Zdravkovic. 2008. Empirical investigation of starling flocks: A benchmark study in collective animal behaviour. *Animal Behaviour* 76:201–215.

Balosser, W. H., and S. M. Russell. 2000. Black-chinned Hummingbird (*Archilochus alexandri*). In *The Birds of North America*, no. 495 (A. Poole and F. Gill, eds.) Philadelphia: The Birds of North America, Inc.

Bartholomew, G. A., T. R. Howell, and T. J. Cade. 1957. Torpidity in the White-throated Swift, Anna Hummingbird, and Poor-will. *Condor* 59(3):145–155.

Barton, D. 1982. Notes on skuas and jaegers in the western Tasman Sea. *Emu* 82:56–59.

Basili, G. D., and S. A. Temple. 1999. Winter ecology, behavior, and conservation needs of Dickcissels in Venezuela. *Studies in Avian Biology* 19:289–299

Batra, T., I. Malik, and V. Kumar. 2019. Illuminated night alters behaviour and negatively affects physiology and metabolism in diurnal zebra finches. *Environmental Pollution* 254(part A):112916.

Baumgartner, A. M. 1937. Food and feeding habits of the Tree Sparrow. *Wilson Bull.* 49(2):65–80.

Beauchamp, G. 1999. The evolution of communal roosting in birds: Origin and secondary losses. *Behavioral Ecology* 10(6):675–687.

———. 2016. Timing of attacks by a predator at a prey hotspot. *Behav. Ecol. Sociobiol.* 70:269–276.

Bechard M. J., and T. R. Swem. 2002. Rough-legged Hawk (*Buteo lagopus*). In *The Birds of North America*, no. 641 (A. Poole and F. Gill, eds.) Philadelphia: The Birds of North America, Inc.

Béchet, A., J-F Giroux, G. Gauthier, and M. Bélisle. 2010. Why roost at the same place? Exploring short-term fidelity in staging Snow Geese. *Condor* 112(2): 294–303.

Beebe, W., G. I. Hartley, and P. G. Howes. 1917. *Tropical Wild Life in British Guiana*. New York: NY Zoological Society.

Beletsky, L. 1996. *The Red-winged Blackbird: The Biology of a Strongly Polygynous Songbird*. London: Academic Press.

Belton, W. 1984. Birds of Rio Grande do Sul, Brazil. Part 1: Rheidae through Furnariidae. *Bull. Am. Mus. Nat. Hist.* 178:371–631.

Berger, R. J., and N. H. Phillips. 1995. Energy conservation and sleep. *Behavioural Brain Research* 69:65–73.

Biebach, H. 1977. Das Winterfett der Amsel (*Turdus merula*). *J. Ornithol.* 118:117–133.

Bijleveld, M. Egas, J. A. van Gils, and T. Piersma. 2010. Beyond the information centre hypothesis: Communal roosting for information on food, predators, travel companions and mates? *Oikos* 119:277–285.

Bijlsma, R. G. 1990. Predation by large falcons on wintering waders on the Banc d'Arguin, Mauritania. *Ardea* 78(1):75–82.

Bijlsma, R. G., and B. van den Brink. 2005. A Barn Swallow *Hirundo rustica* roost under attack: timing and risks in the presence of African Hobbies *Falco cuvieri*. *Ardea* 93(1):36–48.

Bildstein, K. L. 1993. *White Ibis: Wetland Wanderer*. Washington: Smithsonian Institution Press.

Birkhead, T. 1991. *The Magpies*. London: T & A. D. Poyser.

Birkhead, T. R. 1973. A winter roost of Grey Herons. *Brit. Birds* 66(4):147–156.

Bishop, R. P., and A. L. Groves. 1991. The social structure of Arabian babbler, *Turdoides squamiceps*, roosts. *Animal Behaviour* 42(2):323–325.

Black, J. D. 1932. A winter robin roost in Arkansas. *Wilson Bull.* 44(1):13–19.

Blanco, G., and J. L. Tella. 1999. Temporal, spatial and social segregation of red-billed choughs between two types of communal roost: A role for mating and territory acquisition. *Animal Behaviour* 57:1219–1227.

Bobbo, D., F. Galvani, G. G. Mascetti, and G. G. Vallortigara. 2002. Light exposure of the chick embryo influences monocular sleep. *Behavioral Brain Research* 134:447–466.

Bock, A., B. Naef-Daenzer, H. Keil, F. Korner-Nievergelt, M. Perrig, and M. U. Grüebler. Roost site selection by Little Owls *Athene noctua* in relation to environmental conditions and life-history stages. *Ibis* 155:847–856.

Bodrati, A., K. L. Cockle, F. G. Di Sallo, C. Ferreyra, S. A. Salvador, and M. Lammertink. 2015. Nesting and social roosting of the Ochre-collared Piculet (*Picumnus temminckii*) and White-barred Piculet (*Picumnus cirratus*), and implications for the evolution of woodpecker (Picidae) breeding biology. *Orn. Neotrop.* 26(3):223–244.

Boix-Hinzen, C. and B.G. Lovegrove. 1998. Circadian metabolic and thermoregulatory patterns of red-billed woodhoopoes (*Phoeniculus purpureus*): The influence of huddling. *J. Zool., London* 244:33–41.

Bond, A. B. and J. Diamond. 2019. *Thinking like a Parrot: Perspectives from the Wild*. Chicago: U. of Chicago Press.

Bouchard, L. C., and M. J. Anderson. 2011. Caribbean Flamingo resting behavior and the influence of weather variables. *J. Ornithology* 152:307–312.

Boxall, P. C., and M. R. Lein. 1989. Time budgets and activity of wintering Snowy Owls. *J. Field Ornithol.* 60(1):20–29.

Bremond, J. C., P. Gramet, T. Brough, and E. N. Wright. 1968. A comparison of some broadcasting equipments and recorded distress calls for scaring birds. *J. Applied Ecol.* 5(3):521–529.

Brewer, D. 2001. *Wrens, Dippers and Thrashers*. New Haven: Yale U. Press.

Brewster, W. 1890. Summer robin roosts. *Auk* 7:360–373.

Brodsky, L. M. and P. J. Weatherhead. 1984. Behavioural thermoregulation in wintering black ducks: Roosting and resting. *Can. J. Zool.* 62:1223–1226.

Brooke, M. 2004. *Albatrosses and Petrels across the World*. Oxford: Oxford U. Press.

Brooke, R. K. 1965. Roosting of the Black-shouldered Kite *Elanus caeruleus* (Desfontaines). *Ostrich* 36(1):43.

Broom, D. M., W.J.A. Dick, C. E. Johnson, D. I. Sales, and A. Zahavi. 1976. Pied Wagtail roosting and feeding behaviour. *Bird Study* 23(4):267–279.

Brothers, N., R. Gales, and T. Reid. 1999. The influence of environmental variables and mitigation measures on seabird catch rates in the Japanese tuna longline fishery within the Australian Fishing Zone, 1991–1995. *Biol. Cons.* 88:85–101.

Brough, T. 1969. The dispersal of starlings from woodland roosts and the use of bio-acoustics. *J. Applied Ecology* 6(3):403–410.

Brown, C. R. 1980. Sleeping behavior of Purple Martins. *Condor* 82:170–175.

Brown, D. E., J. C. Hagelin, M. Taylor, and J. Galloway. 1998. Gambel's Quail (*Callipepla gambelii*). In *The Birds of North America*, no. 321 (A. Poole and F. Gill, eds.) Philadelphia: The Birds of North America, Inc.

Brown, L. 1980. *The African Fish Eagle*. Capetown: Purnell & Sons.

Brown, L., and D. Amadon. 1968. *Eagles, Hawks and Falcons of the World*. New York: McGraw-Hill.

Brua, R. B. 2001. Ruddy Duck (*Oxyura jamaicensis*). In *The Birds of North America*, no. 696 (A. Poole and F. Gill, eds.) Philadelphia: The Birds of North America, Inc.

Brugger, K. E., R. F. Labinsky, and D. E. Daneke. 1992. Blackbird roost dynamics at Millers Lake, Louisiana: Implications for damage control in rice. *J. Wildlife Management* 56(2):393–398.

Bucher, T. L., and A. Worthington. 1982. Nocturnal hypothermia and oxygen consumption in manakins. *Condor* 84:327–331.

Buckley, F. G. 1968. Behaviour of the Blue-crowned Hanging Parrot *Loriculus galgulus* with comparative notes on the Vernal Hanging Parrot *L. vernalis*. *Ibis* 110:145–164.

Buckley, N. J. 1996. Food finding and the influence of information, local enhancement, and communal roosting on foraging success of North American vultures. *Auk* 113(2):473–488.

———. 1998. Interspecific competition between vultures for preferred roost positions. *Wilson Bull.* 110(1):122–125.

Buckley, P. A., and F. G. Buckley. 2002. Royal Tern (*Sterna maxima*). In *The Birds of North America*, no. 700 (A. Poole and F. Gill, eds.). Philadelphia: The Birds of North America Inc.

Buehler, D. A. 2000. Bald Eagle (*Haliaeetus leucocephalus*). In *The Birds of North America*, no. 506 (A. Poole and F. Gill, eds.). Philadelphia: The Birds of North America Inc.

Burford, D. J., M. S. Miller, and R. S. Lutz. 1994. Sexual behavior of Rio Grande Wild Turkeys at a Roost. *Bird Behavior* 10(1–2):53–54.

Burger, A. E. 1996. Family Chionidae (Sheathbills). In *Handbook of the Birds of the World*, vol. 3 (J. del Hoyo, A. Elliott, and J. Sargatal, eds.), pp. 546–555. Barcelona: Lynx Edicions.

Burger, J., and M. Gochfeld. 1991a. Human activity influence and diurnal and nocturnal foraging of Sanderlings (*Calidris alba*). *Condor* 93:259–265.

———. 1991b. *The Common Tern: Its Breeding Biology and Social Behavior*. New York: Columbia U. Press.

Burton, N.H.K., P. R. Evans, and M. A. Robinson. 1996. Effects on shorebird numbers of disturbance, the loss of a roost site and its replacement by an artificial island at Hartlepool, Cleveland. *Biol. Conserv.* 77:193–201.

Butler, A. W. 1892. Notes on the range and habits of the Carolina Parakeet. *Auk* 9(1):49–56.

Buxton, E.J.M. 1975. High flight of House Martins. *British Birds* 68(7):299–300.

Byrd, G. V., and J. C. Williams. 1993. Red-legged Kittiwake (*Rissa brevirostris*). In *The Birds of North America*, no. 60 (A. Poole and F. Gill, eds.) Philadelphia: Academy of Natural Sciences; Washington, D.C.: American Ornithologists' Union.

Byrkjedal, I., T. Lislevand, and S. Vogler. 2012. Do passerine birds utilize artificial lights to prolong their diurnal activity during winter at northern latitudes? *Ornis Norvegica* 35:37–42.

Caccamise, D. F., and D. W. Morrison. 1986. Avian communal roosting: Implications of diurnal activity centers. *Am. Naturalist* 128(2):191–198.

Caccamise, D. F., L. M. Reed., J. Romanowski, and P. C. Stouffer. 1997. Roosting behavior and group territoriality in American Crows. *Auk* 114(4):628–637.

Calder, W. A. 1993. Rufous Hummingbird (*Selasphorus rufus*). In *The Birds of North America*, no. 53 (A. Poole and F. Gill, eds.) Philadelphia: Academy of Natural Sciences; Washington, D.C.: American Ornithologists' Union.

Calder, W. A., and J. R. King. 1974. Thermal and caloric relations in birds. In *Avian Biology*, vol. 4 (D. S. Farner and J. R. King, eds.), pp. 260–413. New York: Academic Press.

Calf, K., N. Adams, and R. Slotow. 2002. Dominance and huddling behaviour in Bronze Mannikin *Lonchura cucullata* flocks. *Ibis* 144:488–493.

Campbell, B., and E. Lack. 1985. *A Dictionary of Birds*. Vermillion, SD: Buteo Books.

Campbell, S. S., and I. Tobler. 1984. Animal sleep: A review of sleep duration across phylogeny. *Neuroscience & Biobehavioral Reviews* 8:269–300.

Camphuysen, K.C.J. 2007. Where two oceans meet: Distribution and offshore interactions of great-winged petrels *Pterodroma macroptera* and Leach's storm petrels *Oceanodroma leucorhoa* off southern Africa. *J. Ornithol.* 148:333–346.

Carboneras, C. 1992. Family Procellariidae (Petrels and Shearwaters). In *Handbook of the Birds of the World*, vol. 1, pp. 216–257. Barcelona: Lynx Edicions

Cardoso, T.A.L., and D. Zeppelini. 2013. Migratory shorebirds roosting on a roof in Paraíba, Brazil: Response to a new habitat or loss of the natural ones? *Orn. Neotrop.* 24:225–229.

Carere, C., S. Montanino, F. Moreschini, F. Zoratto, F. Chiarotti, D. Santucci, and E. Alleva. 2009. Aerial flocking patterns of wintering starlings, *Sturnus vulgaris*, under different predation risk. *Animal Behaviour* 77:101–107.

Carpenter, J. P., and Y. Wang. 2018. Diurnal space use and nocturnal roost-site selection by male Cerulean Warblers during the breeding season. *J. Field Orn.* 89(1):47–63.

Carter, M. J. 1969. Dusk ascent of Fork-tailed Swifts. *Australian Bird Watcher* 3:168–171.

Chang, Y-H. and L. H. Tang. 2017. Mechanical evidence that flamingos can support their body on one leg with little active muscular force. *Biology Letters* 13:20160948.

Chantler, P. 1995. *Swifts: A Guide to the Swifts and Treeswifts of the World*. New Haven: Yale U. Press.

Charette, M. R., S. Calmé, and F. Pelletier. 2011. Observations of nocturnal feeding in Black Vultures (*Coragyps atratus*). *J. Raptor Res.* 45(3):279–280.

Cheke, R. A. 1971. Temperature rhythms in African montane sunbirds. *Ibis* 113:500–506.

Cheke, R. A., and C. F. Mann. 2001. *Sunbirds: A Guide to the Sunbirds, Flowerpeckers, Spiderhunters and Sugarbirds of the World*. New Haven: Yale U. Press.

Chevallier, D., R. Duponnois, F. Baillon, P. Brossault, J. M. Grégoire, H. Eva, Y. Le Maho, and S. Massemin. 2010. The importance of roosts for Black Storks *Ciconia nigra* wintering in West Africa. *Ardea* 98(1):91–96.

Christe, P., A. Oppliger, and H. Richner. 1994. Ectoparasite affects choice and use of roost sites in the great tit, *Parus major. Anim. Behav.* 47(4):895–898.

———. 1996. Of great tits and fleas: sleep baby sleep. . . . *Anim. Behav.* 52. 1087–1092.

Clark, G. A., Jr. 1973. Unipedal postures in birds. *Bird Banding* 44(1):22–26.

Clarke, R., V. Prakash, W. S. Clark, N. Ramesh, and D. Scott. 1998. World record count of roosting harriers *Circus* in Blackbuck National Park, Velavadar, Gujarat, north-west India. *Forktail* 14:70–71

Clement, P., and R. Hathway. 2000. *Thrushes.* London: Christopher Helm.

Cohn-Haft, M. 1999. Family Nyctibiidae (Potoos). In *Handbook of Birds of the World: Barn Owls to Hummingbirds*, vol. 5 (J. del Hoyo, A. Elliott, and J. Sargatal, eds.), pp. 288–301. Barcelona: Lynx Edicions.

Collar, N. J. 2005. Family Turdidae (Thrushes). In *Handbook of Birds of the World: Cuckoo-shrikes to Thrushes*, vol. 10 (J. del Hoyo, A. Elliott, and J. Sargatal, eds.), pp. 514–807. Barcelona: Lynx Edicions.

Collins, C. T. 1968. The comparative biology of two species of swifts in Trinidad, West Indies. *Bull. Fla. State Museum* 2(5):257–320.

Connelly, F., M. L. Hall, R. D. Johnson, S. Elliot-Kerr, B. R. Dow, J. A. Lesku, and R. A. Mulder. 2022. Urban noise does not affect cognitive performance in wild Australian magpies. *Anim. Behav.* 188:35–44.

Coombs, F. 1976. *The Crows: A Study of the Corvids of Europe.* London: B. T. Batsford Ltd.

Cooper, S. J. 1999. The thermal and energetic significance of cavity roosting in Mountain Chickadees and Juniper Titmice. *Condor* 101:863–866.

Cooper, S. J., and J. A. Gessaman. 2005. Nocturnal hypothermia in seasonally acclimatized Mountain Chickadees and Juniper Titmice. *Condor* 107:151–155.

Cotterman, V., and B. Heinrich. 1993. A large temporary roost of Common Ravens. *Auk* 110(2):395.

Cougill, S., and S. J. Marsden. 1980. Variability in roost size in an *Amazona* parrot: Implications for roost monitoring. *J. Field Ornithol.* 75(1):67–73.

Craighead, J. J., and F. C. Craighead Jr. 1956. *Hawks, Owls and Wildlife.* Harrisburg, Pa.: Stackpole.

Cramp, S. 1980. *Handbook of the Birds of Europe the Middle East and North Africa*, vol. 2. Oxford: Oxford U. Press.

Crenshaw, J. G., and B. R. McClelland. 1989. Bald Eagle use of a communal roost. *Wilson Bull.* 101(4): 626–633.

Croll, D. A., L. T. Balance, B. G. Würsig, and W. B. Tyler. 1986. Movements and daily activity patterns of a Brown Pelican in Central California. *Condor* 88:258–260.

Crossin, R. S. 1967. The breeding biology of the Tufted Jay. *Proc. W. Fdtn. Vert. Zool.* 1(5):264–299.

Csada, R. D., and R. M. Brigham. 1992. Common Poorwill. In *The Birds of North America*, no. 32 (A. Poole and F. Gill, eds.) Philadelphia: Academy of Natural Sciences; Washington, D.C.: American Ornithologists' Union.

Cullen, J. M. 1954. The diurnal rhythm of birds in the arctic summer. *Ibis* 96:31–46.

Curnutt, J. L. 1992. Dynamics of a year-round communal roost of Bald Eagles. *Wilson Bull.* 104(3):536–540.

Da Silva, A., and B. Kempenaers. 2017. Singing from north to south: Latitudinal variation in timing of dawn singing under natural and artificial light conditions. *J. Anim. Ecol.* 86: 1286–1297.

Da Silva, A., J. M. Samplonius, E. Schlicht, M. Valcu, and B. Kempenaers. 2014. Artificial night lighting rather than traffic noise affects the daily timing of dawn and dusk singing in common European songbirds. *Behav. Ecol.* 25(5):1037–1047.

Dave, A. S., and D. Margoliash. 2000. Song replay during sleep and computational rules for sensorimotor vocal learning. *Science* 290 (5492):812–816.

Davies, S.J.J.F. 2002. *Ratites and Tinamous*. Oxford: Oxford U. Press.

Davidar, P., and E. S. Morton. 1993. Living with parasites: Prevalence of a blood parasite and its effect on survivorship in the Purple Martin. *Auk* 110(1):109–116.

Davis, D. 1979. Morning and evening roosts of Turkey Vultures at Malheur Refuge, Oregon. *Western Birds* 10(3):125–130.

Davis, G. J., and J. F. Lussenhop. 1970. Roosting of starlings (*Sturnus vulgaris*): A function of light and time. *Anim. Behav.* 18:362–365.

de Jong, M., J. Q. Ouyang, A. Da Silva, R.H.A. van Grunsven, B. Kempenaers, M. E. Visser, and K. Spoelstra. 2015. Effects of nocturnal illumination on life-history decisions and fitness in two wild songbird species. *Phil. Trans. R. Soc. B.* 370:20140128.

de Juana, E., F. Suárez, and P. G. Ryan. 2004. Family Alaudidae (Larks). In *Handbook of Birds of the World*, vol. 9 (J. del Hoyo, A. Elliott, and D. A. Christie, eds.), pp. 496–601. Barcelona: Lynx Edicions.

Dekker, D., I. Dekker, D. Christie, and R. Ydenberg. 2011. Do staging Semipalmated Sandpipers spend the high-tide period in flight over the ocean to avoid falcon attacks along the shore? *Waterbirds* 34(2):195–201.

Dekker, D., and R. Ydenberg. 2004. Raptor predation on wintering Dunlins in relation to the tidal cycle. *Condor* 106:415–419.

de Lima Moraes, L.J.C.L. 2019. Please, more tears: A case of a moth feeding on antbird tears in central Amazonia. *Ecology* 100(2): e02518.

Dewasmes, G., C. Buchet, A. Geleon, and Y. LeMaho. 1989. Sleep changes in emperor penguins during fasting. *Am. J. of Physiology-Regulatory, Integrative and Comparative Physiology* 256(2):R476–R480.

Dewasmes, G., F. Cohen-Adad, H. Koubi, and Y. LeMaho. 1984. Sleep changes in long-term fasting geese in relation to lipid and protein metabolism. *Am. J. of Physiology-Regulatory, Integrative and Comparative Physiology* 247(4): R663–R671.

———. 1985. Polygraphic and behavioral study of sleep in geese: Existence of nuchal atonia during paradoxical sleep. *Physiology & Behavior* 35(1):67–73.

Dewasmes, G., and N. Loos. 2002. Diurnal sleep depth changes in the king penguin (*Aptenodytes patagonicus*). *Polar Biology* 25:865–867.

Dias, M. P., J. P. Granadeiro, M. Lecoq, C. D. Santos, and J. M. Palmeirim. 2006. Distance to high-tide roosts constrains the use of foraging areas by dunlins: Implications for the management of estuarine wetlands. *Biol. Cons.* 131:446–452.

Dibnah, A. J., J. E. Herbert-Read, N. J. Boogert, G. E. McIvor, J. W. Jolles, and A. Thornton. 2022. Vocally mediated consensus decisions govern mass departures from jackdaw roosts. *Current Biology* 32:455–456.

Dodd, W. L., and M. A. Colwell. 1996. Season variation in diurnal and nocturnal distributions of nonbreeding shorebirds at North Humboldt Bay, California. *Condor* 98:196–207.

Dokter, A. M., S. Åkesson, H. Beekhuis, W. Bouten, L. Buurma, H. van Gasteren, and I. Holleman. 2013. Twilight ascents by common swifts, *Apus apus*, at dawn and dusk: Acquisition of orientation cues? *Animal Behaviour* 85:545–552.

Dominoni, D. M. 2015. The effects of light pollution on biological rhythms of birds: An integrated, mechanistic perspective. *J. Ornithol.* 156 (Suppl. 1):S409–S418.

Dominoni, D. M., E. O. Carmona-Wagner, M. Hofmann, B. Kranstauber, and J. Partecke. 2014. Individual-based measurements of light intensity provide new insights into the effects of artificial light at night on daily rhythms of urban-dwelling songbirds. *J. Anim. Ecol.* 83:681–692.

Döppler, J. F., M. Peltier, A. Amador, F. Goller, and G. B. Mindlen. 2021. Replay of innate vocal patterns during night sleep in suboscines. *Proc. Biol. Sci.* 288(1953):20210610.

Dorst, J. 1957. La vie sur les hauts andins du Pérou. *Revue d'Ecologie, Terre et Vie* 1:3–50.

Dorward, D. F. 1962. Comparative biology of the White Booby and the Brown Booby *Sula* spp. at Ascension. *Ibis* 103b(2): 174–234.

Doucette, D. R., and S. G. Reebs. 1994. Influence of temperature and other factors on the daily roosting times of Mourning Doves in winter. *Can. J. Zool.* 72:1287–1290.

Draulans, D., and J. van Vessem. 1996. Communal roosting in Grey Herons in Belgium. *Colonial Waterbirds* 9:18–24.

Dunn, E. H., and D. J. Agro. 1995. Black Tern (*Chlidonias niger*). In *The Birds of North America*, no. 147 (A. Poole and F. Gill, eds.) Philadelphia: Academy of Natural Sciences; Washington, D.C.: American Ornithologists' Union.

Eckert, A. W. 1974. *The Owls of North America (North of Mexico)*. Garden City, NY: Doubleday & Company, Inc.

Ekner, A., and P. Tryjanowski. 2008. Do small hole nesting passerines detect cues left by a predator? A test on winter roosting sites. *Acta Ornithologica* 43(1):107–111.

Elliott, J. J. 1939. Wintering Tree Swallows at Jones Beach fall and winter of 1937 and 1938. *Bird-Lore* 41:11–16.

Emlen, J. T. 1938. Midwinter distribution of the American Crow in New York state. *Ecology* 19(2):264–275.

Engel, K. A., L. S. Young, K. Steenhof, J. A. Roppe, and M. N. Kochert. 1992. Communal roosting of Common Ravens in southwestern Idaho. *Wilson Bull.* 104(1):105–121.

Erritzoe, J., and H. B. Erritzoe. 1998. *Pittas of the World: A Monograph on the Pitta Family*. Cambridge: Lutterworth Press.

Ewins, P. J., P. D. Rounds, and D. R. Bazely. 1991. Urban roosting by Barn Swallows *Hirundo rustica* wintering in Bangkok. *Forktail* 6:68–70.

Feare, C. 1984. *The Starling*. Oxford: Oxford U. Press.

Feare, C., and A. Craig. 1999. *Starlings and Mynas*. Princeton: Princeton U. Press.

ffrench, R. P. 1967. The Dickcissel on its wintering grounds in Trinidad. *Living Bird* 6:123–140.

Finne, M. H., P. Wegge, S. Eliassen, and M. Odden. 2000. Daytime roosting and habitat preference of capercaillie *Tetrao urogallus* males in spring—the importance of forest structure in relation to anti-predator behaviour. *Wildlife Biology* 6(4):241–249.

Fjeldså, J. 2004. *The Grebes: Podicipedidae.* Oxford: Oxford U. Press.

Forshaw, J. M. 1989. *Parrots of the World: Third (Revised) Edition* Willoughby, Australia: Lansdowne Editions.

———. 2002. *Turacos: A Natural History of the Musiphagidae.* Melbourne: Nokomis Editions Pty Ltd.

Frazier, A., and V. Nolan Jr. 1959. Communal roosting by the Eastern Bluebird in winter. *Bird-Banding* 30:219–226.

Frederick, P. C., and D. Siegel-Causey. 2000. Anhinga (*Anhinga anhinga*). In *The Birds of North America*, no. 522. (A. Poole and F. Gill, eds.) Philadelphia: The Birds of North America, Inc.

French, N. R., and R.W. Hodges. 1959. Torpidity in cave-roosting hummingbirds. *Condor* 61(2):223.

Fry, C. H. 1984. *The Bee-eaters.* Vermillion, SD: Buteo Books.

Fry, C. H., and G. S. Keith. 2000. *The Birds of Africa*, vol. 6. San Diego: Academic Press.

Fry, C. H., S. Keith, and E. K. Urban. 1988. *The Birds of Africa*, vol. 3. London: Academic Press.

Fuller, R. A., P. H. Warren, and K. J. Gaston. 2007. Daytime noise predicts nocturnal singing in urban robins. *Biology. Letters* 3:368–370.

Furness, G., and J.M.C. Peterson. 1987. Common Redpolls excavating snow burrows and snow bathing. *Kingbird* 37(2):74–75.

Furness, R. S. 1987. *The Skuas.* Calton, Staff.: T & A. D. Poyser.

Gadgil, M., and S. Ali. 1975. Communal roosting habits of Indian Birds. *J. Bombay Nat. Hist. Soc.* 72(3): 716–727.

Gallup, A. C. 2022. The causes and consequences of yawning in animal groups. *Anim. Behav.* 187:209–219.

Galton, P. M., and J. D. Shepherd. 2012. Experimental analysis of perching in the European Starling (*Sturnus vulgaris*: Passeriformes; passeres), and the automatic perching mechanism of birds. *J. Experimental Zoology* 317:205–215.

Gao, C., E. M. Morschhauser, D. J. Varricchio, J. Liu, and B. Zhao. 2012. A second soundly sleeping dragon: New anatomical details of the Chinese troodontid *Mei long* with implications for phylogeny and taphonomy. *PLoS One* 7(9):e45203.

García-Walher, J., D. A. Portillo-Zavala, A. Ruiz de Alegria-Arzaburu, and N. R. Senner. 2023. Throwing a lifeline: Floating seagrass rafts as natural alternative roosting habitat for shorebirds. *Ecology* 104(9):c4139.

Gargett, V. 1990. *The Black Eagle.* London: Academic Press.

Gaston, A. 1977. Social behaviour within groups of Jungle Babblers (*Turdoides striatus*). *Animal Behaviour* 25:828–848.

Gauthier, G., and J. Tardif. 1991. Female feeding and male vigilance during nesting in Greater Snow Geese. *Condor* 93:701–711.

Gehlbach, F. R. 1994. *The Eastern Screech Owl: Life History, Ecology, and Behavior in the Suburbs and Countryside.* College Station: Texas A&M U. Press.

Gil, D., M. Honarmand, J. Pascual, E. Pérez-Mena, and C. Macias Garcia. 2015. Birds living near airports advance their dawn chorus and reduce overlap with aircraft noise. *Behav. Ecol.* 26(2):435–443.

Gill, R. E., B. J. McCaffery, and P. S. Tomkovich. 2002. Wandering Tattler (*Heteroscelus incanus*). In *The Birds of North America*, no. 642 (A. Poole and F. Gill, eds.) Philadelphia: The Birds of North America, Inc.

Glahn, J. F., R. D. Flynt, and E. P. Hill. 1994. Historical use of bamboo/cane as blackbird and starling roosting habitat: Implications for roost management. *J. Field Orn.* 65(2):237–246.

Goodwin, D. 1967. *Pigeons and Doves of the World*. London: British Museum (Natural History).

———. 1976. *Crows of the World*. Ithaca, NY: Cornell U. Press.

Gosler, A. G., and P. Clement. 2007. Family Paridae (Tits and Chickadees). In *Handbook of the Birds of the World*, vol. 12. (J. del Hoyo, A. Elliott, and D. A. Christie, eds.) Barcelona: Lynx Edicions.

Gow, E. A., K. L. Wiebe, and J. W. Fox. 2015. Cavity use throughout the annual cycle of a migratory woodpecker revealed by geolocators. *Ibis* 157:167–170.

Gradwohl, J., and R. Greenberg. 1980. The formation of antwren flocks on Barro Colorado Island, Panama. *Auk* 97:385–395.

Graham, J. L., N. J. Cook, K. B. Needham, M. Hau, and T. J. Greives. 2017. Early to rise, early to breed: A role for daily rhythms in seasonal reproduction. *Behavioral Ecology* 28:1266–1271.

Groscolas, R. 1990. Metabolic adaptations to fasting in Emperor and King Penguins. In *Penguin Biology* (L. S. Davis and J. T. Darby eds.), pp. 269–296. San Diego: Academic Press.

Grubb, T. C., Jr. 1982. Downy Woodpecker sexes select different cavity sites: An experiment using artificial snags. *Wilson Bull.* 94(4):577–579.

Guillemette, M. 1994. Digestive-rate constraint in wintering Common Eiders (*Somateria mollissima*): Implications for flying abilities. *Auk* 111(4):900–909.

Gutiérrez, R. J., A. B. Franklin, and W. S. Lahaye. 1995. Spotted Owl (*Strix occidentalis*). In *The Birds of North America*, no. 179 (A. Poole and F. Gill, eds.) Philadelphia: Academy of Natural Sciences; Washington, D.C.: American Ornithologists' Union.

Gwinner, E., and R. Brandstätter. 2001. Complex bird clocks. *Phil. Trans. Royal Soc. London B* 356:1801–1810.

Gyllin, R., H. Källender, and M. Sylvén. 1977. The microclimate explanation of town center roosts of Jackdaws *Corvus monedula*. *Ibis* 119:358–361.

Haftorn, S. 1972. Hypothermia of tits in the arctic winter. *Ornis Scandinavica* 3(2):153–166.

Hailman, J. P., and S. Haftorn. 1995. Siberian Tit (*Parus cinctus*). In *The Birds of North America*, no. 196 (A. Poole and F. Gill, eds.) Philadelphia: Academy of Natural Sciences; Washington, D.C.: American Ornithologists' Union.

Hamilton, L. J., N. K. Michel, J. R. Evanson, and D. L. Roberts. 2022. Surf Scoters use deeper offshore waters during nocturnal resting periods in the Salish Sea of Washington and British Columbia. *Ornithological Applications* 124(3):1–12.

Hancock, J. A., J. A. Kushlan, and M. P. Kahl. 1992. *Storks, Ibises and Spoonbills of the World*. London: Academic Press.

Handel, C. M., and R. E. Gill Jr. 1992. Roosting behavior of premigratory Dunlins (*Calidris alpina*). *Auk* 109(1):57–72.

Haney, J. C. 1986. Seabird patchiness in tropical oceanic waters: The influence of *Sargassum* "reefs." *Auk* 103:141–151.

Hardy, J. W. 1963. Epigamic and reproductive behavior of the Orange-fronted Parakeet. *Condor* 65(3):169–199.

Hardy, S. P., D. R. Hardy, and K. C. Gil. 2018. Avian nesting and roosting on glaciers at high elevation, Cordillera Vilcanota, Peru. *Wilson J. of Ornith.* 130(4):940–957.

Harrap, S. 2008. Family Sittidae (Nuthatches). In *Handbook of the Birds of the World*, vol. 13 (J. del Hoyo, A. Elliott, and D. A. Christie, eds.) Barcelona: Lynx Edicions.

Haug, E. A., B. A. Millsap, and M. S. Martell. 1993. Burrowing Owl (*Speotyto cunicularia*). In *The Birds of North America*, no. 61 (A. Poole and F. Gill, eds.) Philadelphia: The Academy of Natural Sciences; Washington, D.C.: American Ornithologists' Union.

Heath, J. E. 1962. Temperature fluctuation in the Turkey Vulture. *Condor* 64:234–235.

Hedenström, A., G. Norevik, G. Boano, A. Andersson, J. Bäckman, and S. Åkesson. 2019. Flight activity in pallid swifts *Apus pallidus* during the non-breeding period. *J. Avian Biol.* 50(2): e01972.

Hedenström, A., G. Norevik, K. Warfvinge, A. Andersson, J. Bäckman, and S. Åkesson. 2016. Annual 10-month aerial life phase in the Common Swift *Apus apus*. *Current Biology* 26(22):3066–3070.

Hedenström A., R. A. Sparks, G. Norevik, C. Woolley, G. J. Levandoski, and S. Åkesson. 2022. Moonlight drives nocturnal vertical flight dynamics in black swifts. *Current Biology* 32:1875–1881.

Heinrich, B. 1994. Does the early Common Raven get (and show) the meat? *Auk* 111(3):764–769.

———. 2003. Overnighting of Golden-crowned Kinglets during winter. *Wilson Bull.* 115(2):113–114.

Heinsohn, R, G. Olah, M. Webb, R. Peakall, and D. Stojanovic. 2019. Sex ratio bias and shared paternity reduce individual fitness and population viability in a critically endangered parrot. *J. Animal Ecol.* 88:502–510.

Heller, H. C. 1988. Sleep and hypometabolism. *Canadian J. Zool.* 61(1):61–69.

Hendricks, P. 1981. Observations on a winter roost of rosy finches in Montana. *J. Field Orn.* 52(3):235–236.

Henson, P., and J. A. Cooper. 1994. Nocturnal behavior of Trumpeter Swans. *Auk* 111(4): 1013–1018.

Hicklin, P. W. 1987. The migration of shorebirds in the Bay of Fundy. *Wilson Bull.* 99(4): 540–570.

Hiebert, S. 1993. Seasonal changes in body mass and use of torpor in a migratory hummingbird. *Auk* 110(4):787–797.

Hildenbrandt, H., C. Carere, and C. K. Hemelrijk. 2010. Self-organized aerial displays of thousands of starlings: A model. *Behav. Ecol.* 21(6):1349–1359.

Hilgartner, R., M. Raoilison, W. Büttiker, D. C. Lees, and H. W. Krenn. 2007. Malagasy birds as hosts for eye-frequenting moths. *Biology Letters* 3:117–120.

Holt, D. W., and J. L. Petersen. 2000. Northern Pygmy-Owl (*Glaucidium gnoma*). In *The Birds of North America*, no. 494. (A. Poole and F. Gill, eds.) Philadelphia: The Birds of North America, Inc.

Holyoak, D. T. 2001. *Nightjars and Their Allies: The Caprimulgiformes.* Oxford: Oxford U. Press.

Hötker, H. 2000. When do Dunlins spend high tide in flight? *Waterbirds* 23(3):482–485.

Houston, C. S., D. G. Smith, and C. Rohner. Great Horned Owl (*Bubo virginianus*). In *The Birds of North America*, no. 372 (A. Poole and F. Gill, eds.) Philadelphia: The Birds of North America, Inc.

Hutson, H.P.W. 1956. *The Ornithologists' Guide, Especially for Overseas.* London: H. F. & G. Witherby.

Jackson, J. A., and H. R. Ouellet. 2002. Downy Woodpecker (*Picoides pubescens*). In *The Birds of North America*, no. 613. (A. Poole and F. Gill, eds.) Philadelphia: The Birds of North America, Inc.

Jaeger, A., C. J. Feare, R. W. Summers. C. Lebarbenchon, C. S. Larose, and M. Le Corre. 2017. Geolocation reveals year-round at-sea distribution and activity of a superabundant tropical seabird, the Sooty Tern *Onychoprion fuscatus*. *Frontiers in Marine Science* 4:394.

Jaeger, E. C. 1949. Further observations on the hibernation of the Poor-will. *Condor* 51(3): 105–109.

Janicke, T., and N. Chakarov. 2007. Effect of weather conditions on the communal roosting behavior of common ravens *Corvus corax* with unlimited food resources. *J. Ethol.* 25:71–78.

Janousek, W. M., P. P. Marra, and A. M. Kilpatrick. 2014. Avian roosting behavior influences vector-host interactions for West Nile virus hosts. *Parasites and Vectors* 7:399

Jenni, L. 1991. Microclimate of roost sites selected by wintering Bramblings *Fringilla montifringilla*. *Ornis Scandinavica* 22(4):327–334.

―――. 1993. Structure of a Brambling *Fringilla montifringilla* roost according to sex, age and body-mass. *Ibis* 135:85–90.

Jirinec, V., C. P. Varian, C. J. Smith, and M. Leu. 2016. Mismatch between diurnal home ranges and roosting areas in the Wood Thrush (*Hylocichla mustelina*): Possible role of habitat and breeding stage. *Auk* 133(1):1–12.

Johnson, K., and B. D. Peer. 2001. Great-tailed Grackle (*Quiscalus mexicanus*). In *The Birds of North America*, no. 576 (A. Poole and F. Gill, eds.) Philadelphia: The Birds of North America, Inc.

Johnson, M. D., and J. D. Gilardi. 1996. Communal roosting of the Crested Caracara in southern Guatemala. *J. Field Orn.* 67(1):44–47.

Johnson, O. W. and R. M. Nakamura. 1981. The use of roofs by American Golden-Plovers in winter on Oahu, Hawaiian Islands. *Wader Study Group Bull.* 31:45–46.

Jolles, J. A., A. J. King, A. Manica, and A. Thornton. 2013. Heterogeneous structure in mixed-species corvid flocks in flight. *Animal Behaviour* 85:743–750.

Jones, S. G., V. V. Vyazovskiy, C. Cirelli, G. Tononi, and R. M. Benca. 2008. Homeostatic regulation of sleep in the white-crowned sparrow (*Zonotrichia leucophrys gambelii*). *BMC Neuroscience* 9:47

Kavanau, J. L. 1998. Vertebrates that never sleep: Implications for sleep's basic function. *Brain Research Bull.* 46(4):269–279.

Ke, D., and X. Lu. 2009. Burrow use by Tibetan Ground Tits *Pseudopodoces humilis*: Coping with life at high altitudes. *Ibis* 151:321–331.

Kellam, J. S. 2003. Pair bond maintenance in Pileated Woodpeckers at roost sites during autumn. *Wilson Bull.* 115(2):186–192.

Kelly, G. M., and J. P. Thorpe. 1993. A communal roost of Peregrine Falcons and other raptors. *British Birds* 86(2):49–52.

Kemp, A. 1995. *The Hornbills: Bucerotiformes*. Oxford: Oxford U. Press.

Kempenaers, B., P. Borgström, P. Loës, E. Schlicht, and M. Valcu. 2010. Artificial night lighting affects dawn song, extra-pair siring success, and lay date in songbirds. *Current Biology* 20:1735–1739.

Kempenaers, B., and A. A. Dhondt. 1991. Competition between Blue and Great Tits for roosting sites in winter: An aviary experiment. *Ornis Scandinavica* 22(1):73–75.

Kepler, A. K. 1977. Comparative study of the Todies (Todidae): With emphasis on the Puerto Rican Tody (*Todus mexicanus*). Cambridge, MA: Nuttall Ornithological Club.

Kepler, C. B. 1967. Polynesian rat predation on nesting Laysan Albatrosses and other Pacific seabirds. *Auk* 84:426–430.

Kernbach, M. E., V. M. Cassone, T. R. Unnasch, and L. B. Martin. 2020. Broad-spectrum light pollution suppresses melatonin and increases West Nile virus-induced mortality in House Sparrows (*Passer domesticus*). *Condor* 122:1–13.

Kessel, B. 1976. Winter activity patterns of Black-capped Chickadees in interior Alaska. *Wilson Bull.* 88(1):36–61.

Kiis, A. 1986. Timing of roosting in Greenfinches *Carduelis chloris*. *Orn. Scand.* 17(1):80–83.

Kilham, L. 1971. Roosting habits of White-breasted Nuthatches. *Condor* 73:113–114.

———. 1983. *Life History Studies of Woodpeckers of Eastern North America*. Cambridge, MA: Nuttall Ornithological Club.

King, D. T. 1996. Movements of Double-crested Cormorants among winter roosts in the Delta Region of Mississippi. *J. Field Ornithol.* 67(2):205–211.

Kingery, H. E., and C. K. Ghalambor. 2001. Pygmy Nuthatch (*Sitta pygmaea*). In *The Birds of North America*, no. 567 (A. Poole and F. Gill, eds.) Philadelphia: The Birds of North America, Inc.

Kirk, D. A., and M. J. Mossman. 1998. Turkey Vulture (*Cathartes aura*). In *The Birds of North America*, no. 339 (A. Poole and F. Gill, eds.) Philadelphia: The Birds of North America, Inc.

Klaassen, M. 1990. Short note on the possible occurrence of heat stress in roosting waders on the Banc d'Arguin, Mauritania. *Ardea* 78:63–65.

Kluijver, H. N. 1950. Daily routines of the Great Tit, *Parus m. major* L. *Ardea* 38(3/4):99–135.

Knorr, O. 1957. Communal roosting of the Pygmy Nuthatch. *Condor* 59(4):398.

Komar, O. 1992. Communal roosting of the Cave Swallow in El Salvador. *Wilson Bull.* 109(2):332–337.

Krügier, K., R. Prinzinger, and K. L. Schuchmann. 1982. Torpor and metabolism in hummingbirds. *J. Comp. Biochem. Physiol.* 73A(4):679–689.

Kuroda, N. 1961. A note on the pectoral muscles of birds. *Auk* 78(2):261–263.

La, V. T. 2012. Diurnal and nocturnal birds vocalize at night: a review. *Condor* 114(2): 245–257.

Lagerström, M. 1979. Goldcrests *Regulus regulus* roosting in the snow. *Ornis Fennica* 56: 170–172.

Lambertucci, S. A., and A. Ruggiero. 2013. Cliffs used as communal roosts by Andean Condors protect the birds from weather and predators. *PLoS One* 8(6):e67304.

Lammertink, M. 2011. Group roosting in the Grey-and-buff Woodpecker *Hemicircus concretus* involving large numbers of shallow cavities. *Forktail* 27:74–77.

Latta, S. C. 2003. Effects of scaley-leg mite infestations on body condition and site fidelity of migratory warblers in the Dominican Republic. *Auk* 120(3):73–743.

Laughlin, A. J., D. R. Sheldon, D. W. Winkler, and C. M. Taylor. 2014. Behavioral drivers of communal roosting in a songbird: A combined theoretical and empirical approach. *Behavioral Ecology* 25(4):734–743.

Lebbin, D. J., M. G. Harvey, T. C. Lenz, M. J. Andersen, and J. M. Ellis. 2007. Nocturnal migrants foraging at night by artificial lights. *Wilson J. Orn.* 119(3):506–508.

Lendrum, D. W. 1984. Sleeping and vigilance in birds. An experimental study of the barbary dove (*Streptopelia resoria*). *Animal Behaviour* 32:243–248.

Lesku, J. A., and L.M.T. Ly. 2017. Sleep origins: Resting jellyfish are sleeping jellyfish. *Current Biology* 27(19):R1060–1062.

Lesku, J. A., D. Martinez-Gonzalez, and N. C. Rattenborg. 2009. Phylogeny and ontogeny of sleep. In *The Neuroscience of Sleep* (R. Stickgold and M. Walker, eds.), pp. 61–69. San Diego: Academic Press.

Lesku, J. A., L.C.R. Meyer, A. Fuller, S. K. Maloney, G. Dell'Omo et al. 2011. Ostriches sleep like platypuses. *PLoS One* 6(8): e23203.

Lesku, J. A., and N. C. Rattenborg. 2014. Avian sleep. *Current Biology* 24(1):R12-R14.

Lesku, J. A., N. C. Rattenborg, M. Valcu et al. 2012. Adaptive sleep loss in polygynous Pectoral Sandpipers. *Science* 337(6102):1654–1658.

Lesku, J. A., A. L. Vyssotski, D. Martinez-Gonzalez, C. Wilzeck, and N. C. Rattenborg. 2011. Local sleep homeostasis in the avian brain: convergence of sleep functions in mammals and birds? *Proc. R. Soc. B* 278:2419–2428.

Levinson, S. T. 1980. The social behavior of the White-fronted Amazon (*Amazona albifrons*). In *Conservation of New World Parrots: Proceedings of the ICBP Parrot Working Group Meeting, St. Lucia, 1980* (R. F. Pasquier, ed.) Washington, D.C.: Smithsonian Institution Press.

Li, Y., Y. Wan, H. Shen, S. R. Loss, P. P. Marra, and Z. Li. 2021. Estimates of wildlife killed by free-ranging cats in China. *Biol. Cons.* 253:108925.

Liao, W., J. Hu, L. Cao, and X. Lu. 2008. Roosting behaviour of the endangered Sichuan Hill-partridge *Arborophila rufipectus* during the breeding season. *Bird Conservation International* 18(3):260–266.

Libourel, P. A., W. Y. Lee, I. Achin, H. Chung, J. Kim, B. Massot, and N. C. Rattenborg. 2023. Nesting chinstrap penguins accrue large quantities of sleep through seconds-long micro-sleeps. *Science* 382(6674):1026–1031.

Ligon, J. D. 1970. Still more responses of the Poor-will to low temperatures. *Condor* 72(4): 496–498.

Ligon, J. D., and S. Ligon. 1978. The communal social system of the Green Woodhoopoe in Kenya. *Living Bird* 17:159–197.

Lilleyman, A., D. C. Franklin, J. K. Szabo, and M. J. Lawes. 2016. Behavioural responses of migratory shorebirds to disturbance at a high-tide roost. *Emu* 116:111–118.

Lima, S. L. 1988. Initiation and termination of daily feeding in Dark-eyed Juncos: Influences of predation risk and energy reserves. *Oikos* 53(1):3–11.

Lindell, C. 1996. Benefits and costs to Plain-fronted Thornbirds (*Phacellodromus rufifrons*) of interactions with avian nest associates. *Auk* 113(3):565–577.

Lindsey, G. D., E. A. Vanderwerf, H. Baker, and P. E. Baker. 1998. Hawai'i 'Amakihi (*Hemignathus virens*), Kaua'i (*Hemignathus kauaiensis*), O'ahu (*Hemignathus chloris*), and Greater 'Amakihi (*Hemignathus sagittirostris*). In *The Birds of North America*, no. 360 (A. Poole and F. Gill, eds.) Philadelphia: The Birds of North America, Inc.

Löhrl, H. 1955. Schlafgewohnheiten der Baumläufer (*Certhia brachydactyla, C. familiaris*) und anderer Kleinvögel in kalten Winternächten. *Vogelwarte* 18(2):71–77.

Losito, M. P., R. E. Mirarchi, and G. A. Baldassarre. 1990. Summertime activity budgets of hatching-year Mourning Doves. *Auk* 107(1):18–24.

Loss, S. R., T. Will., and P. P. Marra. 2013. The impact of free-ranging domestic cats on wildlife of the United States. *Nature Communications* 4, doi:10.1038/ncomms2380.

Lowther, P. E., and C. L. Cink. 1992. House Sparrow. In *The Birds of North America*, no. 12 (A. Poole, P. Stettenheim, and F. Gill, eds.) Philadelphia: The Academy of Natural Sciences; Washington, D.C.: American Ornithologists' Union.

Maccarone, A. D. 1997. Direction of foraging flights by wading birds during an annular eclipse. *Colonial Waterbirds* 20(3):537–539.

Maclean, G. L. 1973. The Sociable Weaver, part 2: Nest architecture and social organization. *Ostrich* 44:191–218.

MacMillan, R. E., and F. L. Carpenter. 1980. Evening roosting flights of the honeycreepers *Himatione sanguinea* and *Vesteria coccinea* on Hawaii. *Auk* 97:28–37.

Maddocks, T. A., and F. Geiser. 1997. Energetics, thermoregulation and nocturnal hypothermia in Australian Silvereyes. *Condor* 99:104–112.

Mandel, J. T., and K. L. Bildstein. 2007. Turkey Vultures use anthropogenic thermals to extend their daily activity period. *Wilson J. Orn.* 119(1):102–105.

Mane, A. K., and S. S. Manchi. 2017. Roosting patterns of the Edible-Nest Swiftlet (*Aerodramus fuciphagus*) of the Andaman Islands: effects of lunar phase and breeding cycle. *Emu* 117(4):325–332.

Marin, A. M., and F. G. Stiles, 1992. On the biology of fives species of swifts in Costa Rica. *Proc. W. Fdtn. Vert. Zool.* 4(5):285–251.

Marjakangas, A. 1990. A suggested antipredator function for snow-roosting behaviour in the Black Grouse *Tetrao tetrix*. *Ornis Scandinavica* 21(1):77–78.

———. 1992. Winter activity patterns of the Black Grouse. *Ornis Fennica* 69(4):184–192.

Marjakangas, A., H. Rintamäki, and R. Hissa. 1984. Thermal responses in the Capercaillie *Tetrao urogallus* and the Black Grouse *Lyrurus tetrix* roosting in snow. *Physiol. Zool.* 57(1):99–104.

Marks, J. S., T. L. Tibbitts, R. E. Gill, and B. J. McCaffery. 2002. Bristle-thighed Curlew (*Numenius tahitiensis*). In *The Birds of North America*, no. 705 (A. Poole and F. Gill, eds.) Philadelphia: The Birds of North America, Inc.

Marples, B. J. 1934. The winter starling roosts of Great Britain, 1932–33. *J. Anim. Ecol.* 3(2): 187–203.

Marples, G., and A. Marples. 1934. *Sea Terns or Sea Swallows*. London: Country Life Limited.

Martinez-Gonzalez, D., J. A. Lesku, and N. C. Rattenborg. 2008. Increased EEG spectral power density during sleep following short-term sleep deprivation in pigeons (*Columbia livia*): evidence for avian sleep homeostasis. *J. Sleep Research* 17(2):140–153.

Marzluff, J. M., B. Heinrich, and C. S. Marzluff. 1996. Raven roosts as mobile information centers. *Anim. Behav.* 51:89–103.

Mateo-Tomás, P., and P. P. Olea. 2018. Griffon Vultures scavenging at night: trophic niche expansion to reduce intraspecific competition? *Ecology* 99(8):1897–1899.

Matthysen, E. 1998. *The Nuthatches*. London: T & A. D. Poyser.

McCaffery, B., and R. Gill. 2001. Bar-tailed Godwit (*Limosa lapponica*). In *The Birds of North America*, no. 581 (A. Poole and F. Gill, eds.) Philadelphia: The Birds of North America, Inc.

McGowan, A., S. P. Sharp, M. Simeoni, and B.J. Hatchwell. 2006. Competing for position in the communal roosts of long-tailed tits. *Animal Behaviour* 72(5):1035–1043.

McKechnie, A. E., R.A.M. Ashdown, M. B. Christian, and R. M. Brigham. 2007. Torpor in an African caprimulgid, the freckled nightjar *Caprimulgus tristigma*. *J. Avian Biology* 38: 261–266.

McKechnie, A. E., and B. G. Lovegrove. 2001. Thermoregulation and the energetic significance of clustering behavior in the White-backed Mousebird (*Colius colius*). *Physiological and Biochemical Zoology* 74(2):238–249.

———. 2002. Avian facultative hypothermic responses: A review. *Condor* 104:705–724.

McNair, D. B. 1988. Massive roost of Fish Crows at Drum Island, Charleston, South Carolina. *Chat* 52:12–13.

McNaughton, E. J., J. R. Beggs, K. J. Gaston, D. N. Jones, M. C. Stanley. 2021. Retrofitting streetlights with LEDs has limited impact on urban wildlife. *Biological Cons.* 254:108944.

McNeil, R., P. Drapeau, and R. Pierotti. 1993. Nocturnality in colonial waterbirds: Pccurrence, special adaptations, and suspected benefits. In *Current Ornithology*, vol. 10 (D. M. Power, ed.) New York: Plenum Press.

McNicholl, M. K. 1979. Communal roosting of Song Sparrows under snowbank. *Can. Field-Nat.* 93:325–326.

McVey, K. J., P.D.B. Skrade, and T. A. Sordahl. 2008. Use of a communal roost by Turkey Vultures in northeastern Iowa. *J. Field Ornith.* 79(2):170–175.

Mean, C. J., and J. D. Harrison. 1979. Sand Martin movements within Britain and Ireland. *Bird Study* 26(2):73–86.

Meanley, B. 1965. The roosting behavior of Red-winged Blackbirds in the southern United States. *Wilson Bull.* 77(3):217–228.

Meier, C. M., H. Karaadiç, R. Aymí, S. G. Peev, E. Bächler, R. Weber, W. Witvliet, and F. Liechti. 2018. What makes Alpine swift ascent at twilight? Novel geolocators reveal year-round flight behaviour. *Behavioral Ecology and Sociobiology* 72(3), article no. 45.

Merola-Zwartjes, M. 1998. Metabolic rate, temperature regulation, and the energetic implications of roost nests in the Bananaquit (*Coereba flaveola*). *Auk* 115(3):780–786.

Michael, E. D., and W. H. Chao. 1973. Migration and roosting of Chimney Swifts in east Texas. *Auk* 90:100–105.

Miché, F., B. Viven-Roels, P. Pévet, C. Spehner, J. P. Robin, and Y. Le Maho. 1991. Daily pattern of melatonin secretion in an Antarctic bird, the Emperor Penguin, *Aptenodytes forsteri*: Seasonal variations, effect of constant illumination and of administration of isoproterenol or propranolol. *General and Comparative Endocrinology* 84:249–263.

Miller, L. 1936. Some maritime birds observed off San Diego, California. *Condor* 38(1):9–16.

Mitchell, J. R., M. E. Moser, and J. S. Kirby. 1988. Declines in midwinter counts of waders roosting on the Dee estuary. *Bird Study* 35(3):191–198.

Moreau, R. E. 1972. *The Palaearctic-African Bird Migration Systems*. London: Academic Press.

Morris, R. D., and D. A. Wiggins. 1986. Ruddy Turnstones, Great Horned Owls, and egg loss from Common Tern clutches. *Wilson Bull.* 98(1):101–109.

Morton, E. S., and R. M. Patterson. 1983. Kin association, spacing, and composition of a post-breeding roost of Purple Martins. *J. Field Ornith.* 54(1):36–41.

Mowbray, T. B. 2002. Northern Gannet (*Morus bassanus*). In *The Birds of North America*, no. 693 (A. Poole and F. Gill, eds.) Philadelphia: The Birds of North America, Inc.

Mowbray, T. B., F. Cooke, and B. Ganter. 2000. Snow Goose (*Chen caerulescens*). In *The Birds of North America*, no. 514. (A. Poole and F. Gill, eds.) Philadelphia: The Birds of North America, Inc.

Mundy, P., D. Butchart, J. Ledger, and S. Piper. 1992. *The Vultures of Africa*. London: Academic Press.

Munn, C. A., and J. W. Terborgh. 1979. Multi-species territoriality in Neotropical foraging flocks. *Condor* 81:338–347.

Nee, K., and V.Y.Y. Yeo. 1993. Roost site selection and the waking and roosting behaviour of mynas in relation to light intensity. *Malayan Nature J.* 46(3&4):255–263.

Nelson, J. B. 2005. *Pelicans, Cormorants and their relatives: Pelecanidae, Sulidae, Phalacrocoracidae, Anhingidae, Fregatidae, Phaetontidae*. Oxford: Oxford U. Press.

Newton, I. 1973. *Finches*. New York: Taplinger.

Nickley, B., and L. P. Bulluck. 2019. Red-headed Woodpecker (*Melanerpes erythrocephalus*) winter roost-site selection in a burned forest stand. *Wilson J. of Ornith.* 131(4): 774–788.

Nilsson, C., J. Bäckman, and A. M. Dokter. 2019. Flocking behaviour in the twilight ascents of Common Swifts *Apus apus*. *Ibis* 161:674–678.

Nomano, F. Y., 2021. Communal roosting shows dynamics predicted by direct and indirect nepotism in chestnut-crowned babblers. *Behavioral Ecology and Sociobiology* 75:27.

Norris, R. A. 1958. Comparative biosystematics and life history of the nuthatches *Sitta pygmaea* and *Sitta pusilla*. *U. Calif. Pub. in Zool.* 56:119–300.

Novikov, G. A. 1972. The use of under-snow refuges among small birds of the sparrow family. *Aquilo Ser. Zool.* 13:95–97.

Ockendon, N., S. E. Davis, M. P. Toms, and S. Mukherjee. 2009. Eye size and the time of arrival of birds at garden feeding stations in winter. *J. Ornith.* 140:903–908.

Odum, E. P., and F. A. Pitelka. 1939. Storm mortality at a winter starling roost. *Auk* 56:451–455.

Oniki, Y., and E. O. Willis. 2002. Roosting behavior of the Sayaca Tanager (*Thraupis sayaca*) in southeastern Brazil. *Ornith. Neotropical* 13:195–196.

Oren, D. 1980. Enormous concentration of martins (*Progne* spp.) in Iquitos, Peru. *Condor* 82:344–345.

Ouyang, J. Q., M. de Jong, R.H.A. van Grunsven, K. D. Matson, M. F. Haussmann, P. Meerlo, M. E. Visser, and K. Spoelstra. 2017. Restless roosts: Light pollution affects behavior, sleep, and physiology in a free-living songbird. *Global Change Biology* 23:4987–4994.

Paclik, M., and K. Weidinger. 2007. Microclimate of tree cavities during winter nights—implications for roost site selection in birds. *Int. J. of Biometeorology* 51:287–293.

Padró, J., J. N. Pauli, P. L. Perrig, and S. A. Lambertucci. 2019. Genetic consequences of social dynamics in the Andean condor: The role of sex and age. *Behavioral Ecology and Sociobiology* 73:100.

Page, G., and D. F. Whitacre. 1975. Raptor predation on wintering shorebirds. *Condor* 77:73–83.

Palmgren, P. 1949. On the diurnal rhythm of activity and rest in birds. *Ibis* 91(4):561–576.

Palomeque, J., A. Rodriguez, J. D. Palacios, and J. Lanas. 1980. Blood respiratory properties of swifts. *Comp. Biochem. and Physiol.* 67A:91–95.

Parker, P. G., T. A. Waite, and M. D. Decker. 1995. Kinship and association in communally roosting black vultures. *Anim. Behav.* 49:395–401.

Parry, V. A. 1972. *Kookaburras.* New York: Taplinger.

Pearson, O. P. 1953. Use of caves by hummingbirds and other species at high altitudes in Peru. *Condor* 55(1):17–20.

Pendlebury, C. J., and D. M. Bryant. 2005. Night-time behavior of egg-laying tits. *Ibis* 147:343–345.

Perlmutter, G. B. 1992. Environmental factors influencing roost arrival of Black-crowned Night-herons. *J. Field Orn.* 63(4):462–465.

Perrins, C. M. 1979. *British Tits.* London: Collins.

Peters, K. A., and D. L. Otis. 2007. Shorebird roost-site selection at two temporal scales: Is human disturbance a factor? *J. Applied Ecology* 44:196–209.

Pettigrew, J. D., and P. Wilson. 1985. Nocturnal hypothermia in the White-throated Needletail *Hirundapus caudacutus. Emu* 85:200–201.

Pfister, C., B. A. Harrington, and M. Lavine. 1992. The impact of human disturbance on shorebirds at a migration staging area. *Biol. Conserv.* 60:115–126.

Phelan, B., R. A. Phillips, J.R.D. Silk, V. Afanasyev et al. 2007. Foraging behavior of four albatross species by night and day. *Marine Ecology Progress Series* 340:271–286.

Pierotti, R. J., and T. P. Good. 1994. Herring Gull (*Larus argentatus*). In *The Birds of North America,* no. 124 (A. Poole and F. Gill, eds.) Philadelphia: Academy of Natural Sciences; Washington, D.C.: American Ornithologists' Union.

Piersma, T., R. E. Gill Jr., A. de Goeij, A. Dekinga, M. L. Shepherd, M. L. Ruthrauff, and L. Tibbitts. 2006. Shorebird avoidance of nearshore feeding and roosting areas at night correlates with presence of a nocturnal avian predator. *Wader Study Group Bulletin* 109:73–76.

Piersma, T., B. Spaans, and A. Dekinga. 2002. Are shorebirds sometimes forced to roost on water in thick fog? *Wader Study Group Bull.* 97:42–44.

Pitches, A. 2013. Frigatebirds return to nest on Ascension Island. *British Birds* 106(1): 4.

Pitter, E., and M. B. Christiansen. 1997. Behavior of individuals and social interactions of the Red-fronted Macaw *Ara rubrogenys* in the wild during midday rest. *Orn. Neotrop.* 8:133–143.

Pitts, T. D. 1976. Fall and winter roosting habits of Carolina Chickadees. *Wilson Bull.* 88(4):603–610.

Platt, S. G., and T. R. Rainwater. 2018. Unusual diurnal roosting behavior by Turkey Vultures (*Cathartes aura*) during a solar eclipse. *Kingbird* 68(1):15–18.

Playá-Montmany, N., E. González-Medina, J. Cabello-Virgel, M. Parejo et al. 2023. Behavioural and physiological responses to experimental temperature changes in a long-billed and long-legged bird: a role for relative appendage size? *Behav. Ecol. and Sociobiol.* 77:7.

Popper, K. J., E. C. Pelren, and J. A. Crawford. 1996. Summer nocturnal roost sites of Blue Grouse in northeastern Oregon. *Great Basin Naturalist* 56(2):177–179.

Portugal, S. J., C. R. White, P. B. Frappell, J. A. Green, and P. J. Butler. 2019. Impacts of "super-moon" events on the physiology of a wild bird. *Ecology and Evolution* 9:7974–7984.

Post, W. 1967. Fish Crows gathering before going to roost. *Chat* 31(3):75–76.

———. 1982. Why do Grey Kingbirds roost communally? *Bird Behav.* 4:46–49.

Post, W., and K. W. Post. 1987. Roosting behavior of the Yellow-shouldered Blackbird. *Fla. Field Nat.* 15(4):93–105.

Post, W., J. P. Poston, and G. T. Bancroft. 1996. Boat-tailed Grackle (*Quiscalus major*). In *The Birds of North America*, no. 207 (A. Poole and F. Gill, eds.) Philadelphia: Academy of Natural Sciences; Washington, D.C.: American Ornithologists' Union.

Pratt, T. K., and E. W. Stiles. 1983. How long fruit-eating birds stay in the plants where they feed: Implications for seed dispersal. *American Naturalist* 122(6):797–805.

Pravosudov, V. V. and J. R. Lucas. 2000. The costs of being cool: A dynamic model of nocturnal hypothermia by small food-caching birds in winter. *J. Avian Biol.* 31:463–472.

Prinzinger, R., and K. Seidle. 1988. Ontogeny of metabolism, thermoregulation, and torpor in the House Martin *Delichon u. urbica* (L.) and its ecological significance. *Oecologia* 76(2): 302–312.

Prior, K. A., and P. J. Weatherhead. 1991. Turkey Vultures foraging at experimental food patches: a test of information transfer at communal roosts. *Behav. Ecol. and Sociobio.* 28(6): 385–390.

Pulliam, H. R. 1973. On the advantages of flocking. *J. Theoretical Biol.* 38(2):419–422.

Rabenold, P. P. 1987a. Roost attendance and aggression in Black Vultures. *Auk* 104(4):647–653.

———. 1987b. Recruitment to food in black vultures: Evidence for following from communal roosts. *Anim. Behav.* 35:1775–1785.

Ralph, C. J., and S. G. Fancy. 1995. Demography and movements of Apapane and Iiwi in Hawaii. *Condor* 97:729–742.

Rattenborg, N. C. 2006a. Evolution of slow-wave sleep and palliopallial connectivity in mammals and birds: A hypothesis. *Brain Research Bull.* 69:20–29.

———. 2006b. Do birds sleep in flight? *Naturwissenschaften* 93:413–425.

———. 2017. Sleeping on the wing. *Interface Focus* 7:20160082.

Rattenborg, N. C., and D. Martinez-Gonzalez. 2015. Avian versus mammalian sleep: The fruits of comparing apples and oranges. *Current Sleep Medicine Reports* 1:55–63.

Rattenborg, N. C., C. J. Amlaner, and S. L. Lima. 2000. Behavioral, neurophysiological and evolutionary perspectives on unihemispheric sleep. *Neuroscience and Biobehavioral Reviews* 24:817–842.

———. 1999. Facultative control of avian unihemispheric sleep under the risk of predation. *Behavioural Brain Research* 105:163–172.

Rattenborg, N. C., D. Martinez-Gonzalez, and J. A. Lesku. 2009. Avian sleep homeostasis: Convergent evolution of complex brains, cognition and sleep functions in mammals and birds. *Neuroscience & Biobehavioral Reviews* 33(3):253–270.

Rattenborg, N. C., D. Martinez-Gonzalez, T. C. Roth II, and V. V. Pravosudov. 2010. Hippocampal memory consolidation during sleep: a comparison of mammals and birds. *Biological Reviews* 86(3):658–691.

Rattenborg, N. C., B. Viorin, S. M. Cruz, R. Tisdale, G. Dell'Omo, H. P. Lipp, M. Wikelski, and A. L. Vyssotski. 2016. Evidence that birds sleep in mid-flight. *Nat. Comm.* DOI:10.1038/ncomms12468.

Rattenborg, N. C., and G. Ungurean. 2023. The evolution and diversification of sleep. *Trends in Ecology and Evolution* 38(2):156–170

Raveling, D. G., W. E. Crews, and W. K. Klimstra. 1972. Activity patterns of Canada Geese during winter. *Wilson Bull.* 84(3):278–295.

Razack, A., and R. M. Naik. 1965. Studies on the House Swift, *Apus affinis* (G. E. Gray) 3: Awakening and roosting during the nonbreeding period. *Pavo* 3(1):55–71.

Reebs, S. G. 1986. Influence of temperature and other factors on the daily roosting times of black-billed magpies. *Can. J. Zool.* 64(8):1614–1619.

Reidy, J. L., M. M. Stake, and F. R. Thompson, III. 2009. Nocturnal predation of females on nests: An important source of mortality for Golden-cheeked Warblers? *Wilson J. Orn.* 121(2):416–421.

Reierth, R., T. J. Van't Hof, and K. A. Stokkan. 1999. Seasonal and daily variations in plasma melatonin in the High-Arctic Svalbard Ptarmigan (*Lagopus mutus hyperboreus*). *J. Biological Rhythms* 14(4):314–319.

Reinertsen, R. E. 1983. Nocturnal hypothermia and its energetic significance for small birds living in the Arctic and subarctic regions: A review. *Polar Research* 1(3):269–284.

Reinertsen, R. E., S. Haftorn, and E. Thaler. 1988. Is hypothermia necessary for the winter survival of the Goldcrest *Regulus regulus*? *J. Orn.* 129:433–437.

Rheinwald, G. 1975. Übernachten auch Mehlschwalben in der Luft? *Die Vogelwelt* 96(6):221–223.

Rico, D., and L. Sandoval. 2014. Non-random orientation in woodpecker cavity entrances in a tropical rain forest. *Ornithologia Neotropical* 25: 237–243.

Robertson, P. B., and A. F. Schnapf. 1987. Pyramiding behavior in the Inca Dove: Adaptive aspects of day-night differences. *Condor* 89:185–187.

Robertson, R. J., B. J. Stutchbury, and R. R. Cohen. 1992. Tree Swallow. In *The Birds of North America*, no. 11 (A. Poole, P. Stettenheim, and F. Gill, eds.) Philadelphia: Academy of Natural Sciences; Washington, D.C.: American Ornithologists' Union.

Rodriguez, A. B., M. P. Terrón, J. Durán, E. Ortega, and C. Barriga. 2001. Physiological concentrations of melatonin and corticosterone affect phagocytosis and oxidative metabolism of ring dove heterophils. *J. Pineal Res.* 31:31–38.

Rogers, D. I., P. F. Battley, T. Piersma, J. A. van Gils, and K. G. Rogers. 2006a. High-tide habitat choice: Insights from modelling roost selection by shorebirds around a tropical bay. *Animal Behaviour* 72(3):563–575.

Rogers, D. I., T. Piersma, and C. J. Hassell. 2006b. Roost availability may constrain shorebird distribution. *Biological Conservation* 133(2):225–235.

Rogers, L. J., G. Vallortigara, and R. J. Andrew. 2013. *Divided Brains: The Biology and Behaviour of Brain Asymmetries*. New York: Cambridge U. Press.

Rohner, C., C. J. Krebs, D. B. Hunter, and D. C. Currie. 2000. Roost site selection of Great Horned Owls in relation to black fly activity: an anti-parasite behavior? *Condor* 102:950–955.

Roth, T. C. II, J. A. Lesku, C. J. Amlaner, and S. L. Lima. 2006. A phylogenetic analysis of the correlates of sleep in birds. *J. Sleep Res.* 15:395–402.

Roth, T. C. II, and S. L. Lima. 2007. The predatory behavior of wintering *Accipiter* hawks: Temporal patterns in activity of predators and prey. *Oecologia* 152:169–178.

Roth, T. C. II, N. C. Rattenborg, and V. V. Pravosudov. 2010. The ecological relevance of sleep: The trade-off between sleep, memory and energy conservation. *Phil. Trans. R. Soc. B* 365:945–949.

Rudebeck, G. 1955. Some observations at a roost of European Swallows and other birds in the south-eastern Transvaal. *Ibis* 97:572–580.

Ruiz, G. M., P. G. Connors, S. E. Griffin, and F. A. Pitelka. 1989. Structure of a wintering Dunlin population. *Condor* 91:562–570.

Runwal, P. 2020. Repurposed oyster farm bags offer new real estate for migratory birds. *Scientific American* [March 4, 2020].

Rusch, D. H., S. DeStefano, M. C. Reynolds, and D. Lauten. 2000. Ruffed Grouse (*Bonasa umbellus*). In *The Birds of North America*, no. 515 (A. Poole and F. Gill, eds.) Philadelphia: The Birds of North America, Inc.

Russell, K. R., and S. A. Gauthreaux Jr. 1999. Spatial and temporal dynamics of a Purple Martin pre-migratory roost. *Wilson Bull.* 111(3):354–362.

Ryeland, J., M. A. Weston, and M.R.E. Symonds. 2019. Leg length and temperature determine the use of unipedal roosting in birds. *J. Avian Biol.* e02008.

Sanders, F. J., M. C. Handmaker, A. S. Johnson, and N. R. Senner. 2021. Nocturnal roost on South Carolina coast supports nearly half of Atlantic coast population of Hudsonian Whimbrel *Numenius hudsonicus* during northward migration. *Wader Study* 128(2):117–124.

Santos, C. D., A. C. Miranda, J. P. Granadeiro, P. M. Lourenço, S. Saraiva, and J. M. Palmeirim. 2010. Effects of artificial illumination on the nocturnal foraging of waders. *Acta Oecologia* 36:166–172.

Sauer, E.G.F., and E. M. Sauer. 1967. Yawning and other maintenance activities in the South African Ostrich. *Auk* 84(4):571–587.

Sazima, I. 2015. Save your tears: Eye secretions of a Ringed Kingfisher fed upon by an erebid moth. *Revista Brasileira de Ornithologia* 23(4):392–394.

Scardamaglia, R. C., A. Kacelnik, and J. C. Reboreda. 1918. Roosting behavior is related to reproductive strategy in brood parasitic cowbirds. *Ibis* 160:779–789.

Schlicht, L., M. Valcu, P. Loës, A. Girg, and B. Kempenaers. 2014. No relationship between female emergence time from the roosting place and extrapair paternity. *Behavioral Ecology* 25(3):650–659.

Schmidt, M. H. 2014. The energy allocation function of sleep: A unifying theory of sleep, torpor, and continued wakefulness. *Neuroscience & Biobehavioral Reviews* 47:122–153.

Schnell, G. D. 1969. Communal roosts of wintering Rough-legged Hawks (*Buteo lagopus*). *Auk* 86:682–690.

Schodde, R. 1982. *The Fairy-Wrens: A Monograph of the Maluridae*. Melbourne: Lansdowne Editions.

Schorger, A. W. 1938. Unpublished manuscripts by Cotton Mather on the Passenger Pigeon. *Auk* 55:471–477.

———. 1955. *The Passenger Pigeon: Its Natural History and Extinction*. Madison: U. Wisc. Press.

Schreiber, E. A., C. J. Feare, B. A. Harrington, B. G. Murray Jr., W. B. Robertson Jr., M. J. Robertson, and G. E. Woolfenden. 2002. Sooty Tern (*Sterna fuscata*). In *The Birds of North America*, no. 665 (A. Poole and F. Gill, eds.) Philadelphia: The Birds of North America, Inc.

Schreiber, R. W. 1967. Roosting behavior of the Herring Gull in central Maine. *Wilson Bull.* 79(4):421–431.

Schreiber, R. W., and J. L. Chovan. 1986. Roosting by pelagic seabirds: Energetic, populational, and social considerations. *Condor* 88:487–492.

Schwilch, R., T. Piersma, N.A.M. Holmgren, and L. Jenni. 2002. Do migratory birds need a nap after a long non-stop flight? *Ardea* 90(1):149–154.

Seibert, H. C. 1951. Light intensity and the roosting flight of herons in New Jersey. *Auk* 68(1):63–74.

Serventy, D. L. 1970. Torpidity in the White-backed Swallow. *Emu* 70:27–28.

Shaffery, J. P., C. J. Amlaner Jr., N. J. Ball, and M. R. Opp. 1985b. Ecological and behavioral correlates of sleep in free-ranging birds. *Sleep Research* 14:103

Shaffery, J. P., N. J. Ball, and C. J. Amlaner Jr. 1985a. Manipulating daytime sleep in herring gulls (*Larus argentatus*). *Animal Behaviour* 33(2):566–572.

Shamoun-Baranes, J., W. Bouten, C. J. Camphuysen, and E. Baaij. 2011. Riding the tide: Intriguing observations of gulls resting at sea during breeding. *Ibis* 153:411–415.

Shankar, A., R. J. Schroeder, S. M. Wethington, C. H. Graham, and D. R. Powers. 2020. Hummingbird torpor in context: Duration, more than temperature, is the key to nighttime energy savings. *J. Avian Biology* e02305.

Short, L. L. 1973. Habits of some Asian woodpeckers (Aves, Picidae). *Bull. Am. Mus. Nat. Hist.* 152(5).

Short, L., and J. Horne. 2001. *Toucans, Barbets and Honeyguides: Ramphastidae, Capitonidae and Indicatoridae*. Oxford: Oxford U. Press.

Sick, H., and D. M. Teixeira. 1981. Nocturnal activities of Brazilian hummingbirds and flycatchers at artificial illumination. *Auk* 98:191–192.

Siegfried, W. R. 1971. Communal roosting of the Cattle Egret. *Trans. Royal Soc. S. Africa* 39(IV):419–443.

Siegfried, W. R., P.G.H. Frost, I. J. Ball, and D. F. McKinney. 1977. Evening gatherings and night roosting of African Black Ducks. *Ostrich* 48:5–16.

Sitters, H. P., P. M. González, T. Piersma, A. J. Bakker, and D. J. Price. 2001. Day and night feeding habitat of Red Knots in Patagonia: profitability versus safety. *J. Field Orn.* 72(1):86–95.

Simons, T. R. 1985. Biology and behavior of the endangered Hawaiian Dark-rumped Petrel. *Condor* 87:229–245.

Skokkan, K. A. 1992. Energetics and adaptations to cold in ptarmigan in winter. *Ornis Scandinavica* 23:366–370.

Skutch, A. F. 1940. Social and sleeping habits of Central American wrens. *Auk* 57(3):293–312.

———. 1958. Roosting and nesting habits of Araçari toucans. *Condor* 60(4):201–219.

———. 1960. *Life Histories of Central American Birds, II*. Berkeley: Cooper Ornithological Society.

———. 1961. The nest as a dormitory. *Ibis* 103a:50–70.

———. 1973. *The Life of the Hummingbird*. New York: Crown.

———. 1985. *Life of the Woodpecker*. Santa Monica, CA.: Ibis Publishing Co.

———. 1989a. *Birds Asleep*. Austin: U. of Texas Press.

———. 1989b. *Life of the Tanager*. Ithaca, NY: Cornell U. Press.

———. 1991. *Life of the Pigeon*. Ithaca, NY: Cornell U. Press.

———. 1996. *Antbirds and Ovenbirds: Their Lives and Homes*. Austin: U. of Texas Press.

———. 1997. *Life of the Flycatcher*. Norman: U. of Oklahoma Press.

Smith, J.A.M., L. R. Reitsma, L. L. Rockwood, and P. P. Marra. 2008. Roosting behavior of a Neotropical migrant songbird, the northern waterthrush *Seiurus noveboracensis*, during the non-breeding season. *J. Avian Biol.* 39:460–465.

Smith, S. M. 1991. *The Black-capped Chickadee: Behavioral Ecology and Natural History*. Ithaca, NY: Comstock Publishing Assoc.

Snow, D. 1982. *The Cotingas: Bellbirds, Umbrellabirds and Their Allies*. London: British Museum (Natural History).

Snyder, N.F.R., and K. Russell. 2002. Carolina Parakeet (*Conuropsis carolinensis*). In *The Birds of North America*, no. 667 (A. Poole and F. Gill, eds.) Philadelphia: The Birds of North America, Inc.

Solano-Ugalde, A., and A. Arcos-Torres. 2008. Nocturnal foraging observations of the Blue-crowned Motmot (*Momotus momota*) in San José, Costa Rica. *Wilson J. Orn.* 120(3): 653–654.

Sonerud, G. A., H. Hansen, and C. A. Smedshaug. 2002. Individual roosting strategies in a flock-living bird: Movement and social cohesion of hooded crows (*Corvus corone cornix*) from pre-roost gatherings to roost sites. *Behav. Ecol. Sociobiol.* 51:309–318.

Sparling, D. W., and G.L. Krapu. 1994. Communal roosting and foraging behavior of staging Sandhill Cranes. *Wilson Bull.* 106(1):62–77.

Speights, J. R., and W.C. Conway. 2009. Wintering Yellow-bellied Sapsucker time activity budgets in East Texas bottomland hardwood forests. *Wilson J. of Orn.* 121(3):593–599.

Spencer, K. G. 1966. Some notes on the roosting behaviour of starlings. *Naturalist* 898:73–80.

Spiegel, O., W. M. Getz, and R. Nathan. 2013. Factors influencing foraging search efficiency: Why do scarce Lappet-faced Vultures outperform ubiquitous White-backed Vultures? *Am. Naturalist* 181(5):E102-E115.

Sprunt, A., Jr., and E. B. Chamberlain. 1970. *South Carolina Bird Life*. Columbia: U. of South Carolina Press.

Stager, K. E. 1965. An exposed nocturnal roost of migrant Vaux Swifts. *Condor* 67:81–82.

Stahel, C. D., D. Megirian, and S. C. Nicol. 1984. Sleep and metabolic rate in the little penguin, *Eudyptula minor. J. Comp. Physiol. B.* 154:487–494.

Stake, J. D., and P.E. Stake. 1983. Apparent torpidity in Tree Swallows. *Connecticut Warbler* 3(3):36–37.

Stalmaster, M. V., and J. A. Gessaman. 1984. Ecological energetics and foraging behavior of overwintering Bald Eagles. *Ecological Monographs* 54(4):407–428.

Stanback, M. T. 1998. Getting stuck: a cost of communal cavity roosting. *Wilson Bull.* 110(3):421–423.

Stanley, C. Q., M. H. Walter, M. X. Venkatraman, and G. S. Wilkinson. 2016. Insect noise avoidance in the dawn chorus of Neotropical birds. *Anim. Behav.* 112:255–265.

Steiger, S. S., M. Valcu, K. Speolstrra, B. Helm, M. Wikelski, and B. Kempenaers. 2013. When the sun never sets: Diverse activity rhythms under continuous daylight in free-living arctic-breeding birds. *Proc. Roy. Soc. B* 280(1764):1–10.

Stewart, P. A. 1978. Behavioral interactions and niche separation in Black and Turkey Vultures. *Living Bird* 17:79–84.

Stiehl, R. B. 1981. Observations of a large roost of Common Ravens. *Condor* 83:78.

Steinen, E.W.M., A. Brenninkmeijer, and M. Klaasen. 2008. Why do Gull-billed Terns *Gelochelidon nilotica* feed on fiddler crabs *Uca tangeri* in Guinea-Bissau? *Ardea* 96(2):243–250.

Steinmeyer, C., H. Schielzeth, J. C. Mueller, and B. Kempenaers. 2010. Variation in sleep behavior in free-living blue tits, *Cyanestes caeruleus*: Effects of sex, age and environment. *Animal Behaviour* 80(5):853–864.

Still, E., P. Monaghan, and E. Bignal. 1987. Social structuring at a communal roost of Choughs *Pyrrhocorax pyrrhocorax. Ibis* 129:398–403.

Stewart, P. A. 1978. Behavioral interactions and niche separation in Black and Turkey Vultures. *Living Bird* 17:79–84.

Stolen, E. D., and W. K. Taylor. 2003. Movements of Black Vultures between communal roosts in Florida. *Wilson Bull.* 115(3):316–320.

Stromberg, M. R. 2000. Montezuma Quail (*Cyrtonyx montezumae*). In *The Birds of North America*, no. 524 (A. Poole and F. Gill, eds.) Philadelphia: The Birds of North America, Inc.

Stuber, E. F., N. J. Dingemanse, and J. C. Mueller. 2017. Temperature affects frequency but not rhythmicity of nocturnal awakenings in free-living Great Tits, *Parus major. Animal Behaviour* 128:135–141.

Sulkava, S. 1969. On small birds spending the night in the snow. *Aquilo, Ser. Zoologica* 7:33–37.

Summers, R. W., G. E. Westlake, and C. J. Feare. 1987. Differences in the ages, sexes and physical condition of Starlings *Sturnus vulgaris* at the centre and periphery of roosts. *Ibis* 129:96–102.

Summers-Smith, J. D. 1988. *The Sparrows: A Study of the Genus* Passer. Calton, UK:. & A. D. Poyser.

Swengel, S. R., and A. B. Swengel. 1992. Roosts of Northern Saw-whet Owls in southern Wisconsin. *Condor* 94:699–706.

Swennen, C. 1984. Differences in quality of roosting flocks of Oystercatchers. In *Coastal Waders and Wildfowl in Winter* (P. R. Evans, J. D. Goss-Custard, and W. G. Hale, eds.), pp. 177–189. Cambridge: Cambridge U. Press.

Swenson, J. E., and B. Olsson. 1991. Hazel Grouse night roost preferences when snow-roosting is not possible in winter. *Ornis Scandinavica* 22(3):284–286.

Swingland, I. R. 1977. The social organization of winter communal roosting in Rooks (*Corvus frugilegus*). J. Zool. (London) 182:509–528.

Szymczak, J.T., H. W. Helb, and W. Kaiser. 1993. Electrophysiological and behavioral correlates of sleep in the blackbird (*Turdus merula*). *Physiology & Behavior* 53(6):1201–1210.

Sykes, P. W., Jr., J. A. Rodgers Jr., and R. E. Bennett. 1995. Snail Kite (*Rostrhamus sociabilis*). In *The Birds of North America*, no. 171 (A. Poole and F. Gill, eds.) Philadelphia: Academy of Natural Sciences; Washington, D.C.: American Ornithologists' Union.

Tacha, T.C., S. A. Nesbitt, and P. A. Vohs. 1992. Sandhill Crane. In *The Birds of North America*, no. 31 (A. Poole, P. Stettenheim, and F. Gill, eds.) Philadelphia: The Academy of Natural Sciences; Washington, D.C.: American Ornithologists' Union.

Takahishi, J. S., and M. Menaker. 1980. On the organization of avian circadian systems: the role of the pineal and suprachiasmatic nuclei. In *Acta XVII Congressus Internationalis Ornithologici* (R. Nöhring, ed.), pp. 425–434.

Tanner, J. T. 1942. *The Ivory-billed Woodpecker*. New York: National Audubon Society.

Taylor, G. C. 1862. Five weeks in the peninsula of Florida during the spring of 1861, with notes on the birds observed there. *Ibis*:4:127–142.

Taylor, I. 1994. *Barn Owls: Predator-prey relationships and conservation*. Cambridge: Cambridge U. Press.

Taylor, P. B. 1996. Family Rallidae (Rails, Gallinules and Coots). In *Handbook of the Birds of the World*, vol. 3. (J. del Hoyo, A. Elliott, and J. Sargatal, eds.), pp. 108–152. Barcelona: Lynx Edicions

Taylor, R. H. 1985. Status, habits and conservation of *Cyanoramphus* parakeets in the New Zealand region. In *Conservation of Island Birds, ICBP Tech. Publ.*, no. 3: 195–211.

Thomas, R. J., T. Székely, I. C. Cuthill, D.G.C. Harper, S.E. Newson, T. D. Frayling, and P. D. Wallis. 2002. Eye size in birds and the timing of song at dawn. *Proc. R. Soc. Lond. B* 269:831–837.

Thompson, F. R. III and E. K. Fritzell. 1988. Ruffed Grouse winter roost site preference and influence on energy demands. *J. Wildlife Management* 52(3):454–460.

Thomson, A. L. 1964. *A New Dictionary of Birds*. London: Nelson.

Thurber, W. A., and O. Komar. 2002. Turquoise-browed Motmot (*Eumomota superciliosa*) feeds by artificial light. *Wilson J. Orn.* 114(4):525–526.

Tilgar, V., K. Hein, and R. Viigipuu. 2022. Anthropogenic noise alters the perception of a predator in a local community of great tits. *Anim. Behav.* 189:91–99.

Tinbergen, N. 1953. *The Herring Gull's World*. London: Collins.

Tirpak, J. M., W. M. Giuliano, and C. A. Miller. 2005. Nocturnal roost habitat selection by Ruffed Grouse broods. *J. Field Ornith.* 76(2):168–174.

Tobler, I., and A. A. Borbély. 1988. Sleep and EEG spectra in the pigeon (*Columbia livia*) under baseline conditions and after sleep deprivation. *J. Comp. Physiol. A*: 163:729–738.

Tomback, D. F. 1978. Pre-roosting flight of the Clark's Nutcracker. *Auk* 95:554–562.

Tosh, F. E., I. L. Doto, S. B. Beecher, T.D.Y. Chin. 1970. Relationship of starling-blackbird roosts and endemic histoplasmores. *Am. Rev. Respiratory Disease* 101:283–288.

Townsend, J. M., C. C. Rimmer, J. Brocca, K. P. McFarland, and A. K. Townsend. 2009. Predation of a wintering migratory songbird by introduced rats: Can nocturnal roosting behavior serve as predator avoidance? *Condor* 111(3):564–569.

Tramer, E. J. 2000. Bird behavior during a total solar eclipse. *Wilson Bull.* 112(3):431–432.

Troy, D. M. 1983. Recaptures of redpolls: movements of an irruptive species. *J. Field Ornithol.* 54(2):146–151.

Tsuji, A. 1996. *Hornbills: Masters of Tropical Forests.* Bangkok: Sarakadee Press.

Tsutsui, K., S. Haraguchi, K. Inoue et al. 2012. Control of circadian activity of birds by the interaction of melatonin with 7α-hydroxypregnenolone, a newly discovered neurosteroid stimulating locomotion. *J. Ornithology* 153(Suppl. 1):S235-S243.

Ungurean, G., D. Martinez-Gonzalez, B. Massot, P. A. Libourel, and N. C. Rattenborg. 2021. Pupillary behavior during wakefulness, non-REM sleep, and REM sleep in birds is opposite that of mammals. *Current Biology* 31(23):5370–5376.

van den Brink, B., R. G. Bijlsma, and T. M. van der Have. 2000. European Swallows *Hirundo rustica* in Botswana during three non-breeding seasons: the effects of rainfall on moult. *Ostrich* 71(1–2):198–204.

van Hasselt, S. J., R. A. Hut, G. Allocca, A. L. Vyssotski, T. Piersma, N. C. Rattenborg, and P. Meerlo. 2021. Cloud cover amplifies the sleep-suppressing effect of artificial night at night in geese. *Env. Pollution* 273:116444.

van Hasselt, S. J., G. J. Mekenkamp, J. Komdeur, G. Allocca, A.L. Vyssotski, T. Piersma, N. C. Rattenborg, and P. Meerlo. 2020. Seasonal variation in sleep homeostasis in migratory geese: A rebound of NREM sleep following sleep deprivation in summer but not in winter. *Sleep* DOI:10.1093/sleep/zsaa244.

van Hasselt, S. J., M. Rusche, A. L. Vyssotski, S. Verhulst, N. C. Rattenborg, and P. Meerlo. 2020. Sleep time in the European Starling is strongly affected by night length and moon phase. *Current Biology* 30(9):1664–1671.

Van Someren, V.G.L. 1956. *Days with Birds: Studies of Habits of Some East African Species.* Fieldiana: Zoology, vol. 38. Chicago: Chicago Natural History Museum Press.

Velkyu, M., P. Kanuch, and A. Kristin. 2010. Selection of winter roosts in the Great Tit *Parus major*: Influence of microclimate. *J. Ornithol.* 151:147–153.

Verbeek, N.A.M., and C. Caffrey. 2002. American Crow (*Corvus brachyrhynchos*). In *The Birds of North America*, no. 647 (A. Poole and F. Gill, eds.) Philadelphia: The Birds of North America, Inc.

Villard, P. 1999. *The Guadeloupe Woodpecker.* Clichy, France: Société d'Études Ornithologiques de France

Vleck, C. M., and J.A. Van Hook. 2002. Absence of daily rhythms of prolactin and corticosterone in Adélie Penguins under continuous daylight. *Condor* 104:667–671.

Voous, K. H. 1988. *Owls of the Northern Hemisphere.* Cambridge, MA: MIT Press.

Vorster, A. P., and J. Born. 2015. Sleep and memory in mammals, birds and invertebrates. *Neuroscience and Biobehavioral Reviews* 50:103–119.

Waite, T. A. 1991. Nocturnal hypothermia in Gray Jays *Perisoreus canadensis* wintering in interior Alaska. *Ornis Scandinavica* 22(2):107–110.

Wallace, A. R. 1869. *The Malay Archipelago.* New York: Harper & Brothers.

Walsberg, G. E. 1986. Thermal consequences of roost-site selection: The relative importance of three modes of heat conservation. *Auk* 103(1):1–7.

———. 1990. Communal roosting in a very small bird: consequences for the thermal and respiratory gas environments. *Condor* 92:795–798.

Walsberg, G. E., and J. R. King. 1980. The thermoregulatory significance of the winter roost-sites selected by robins in eastern Washington. *Wilson Bull.* 92(1):33–39.

Walters, E. L., E. H. Miller, and P. E. Lowther. 2002. Yellow-bellied Sapsucker (*Sphyrapicus varius*). In *The Birds of North America*, no. 662 (A. Poole and F. Gill, eds.) Philadelphia: The Birds of North America, Inc.

Walther, B. A. 2000. Fruit size and frugivore species richness: additional evidence from observations at a large *Ficus* tree. *Ecotropica* 6:197–201.

———. 2002. Vertical stratification and use of vegetation and light habitats by Neotropical forest birds. *J. Ornithol.* 143:64–81.

Wanless, R. M., A. Angel, R. J. Cuthbert, G. M. Hilton, and P.G. Ryan. 2007. Can predation by invasive mice drive seabird extinctions? *Biology Letters* 3:241–244.

Ward, M. P., M. Alessi, T. J. Benson, and S. J. Chiavacci. 2014. The active nightlife of diurnal birds: Extrapair forays and nocturnal activity patterns. *Animal Behaviour* 88:175–184.

Ward, P. 1965. Feeding ecology of the Black-faced Dioch *Quelea quelea* in Nigeria. *Ibis* 107:173–214.

Ward, P., and A. Zahavi. 1973. The importance of certain assemblages of birds as "information-centres" for food-finding. *Ibis* 115:517–534.

Warham, J. 1957. Notes on the roosting habits of some Australian birds. *Emu* 57:78–81.

———. 1961. The birds of Raine Island, Pandora Cay and Murray Island Sandbank, North Queensland. *Emu* 61(2):76–93.

Warkentin, I. G. 1986. Factors affecting roost departure and entry by wintering merlins. *Can. J. Zool.* 64:1317–1319.

Warkentin, I. G., and E. S. Morton. 1995. Roosting behavior of Prothonotary Warblers in the non-breeding season. *Wilson Bull.* 107(2):374–376.

Warner, R. E. 1968. The role of introduced diseases in the extinction of the endemic Hawaiian avifauna. *Condor* 70:101–120.

Watson, D. 1977. *The Hen Harrier.* Berkhamsted: T & A. D. Poyser.

Watts, B. D., F. M. Smith, C. Hines, et al. 2021. The costs of using night roosts for migrating whimbrels. *J. Avian Biol.* 52(1):e02629

Weatherhead, P. J., and D. J. Hoysak. 1984. Dominance structuring of a Red-winged Blackbird roost. *Auk* 101:551–555.

Weatherhead, P. J., S. G. Sealy, and R.M.R. Barclay. 1985. Risks of clustering in thermally-stressed swallows. *Condor* 87:443–444.

Webb, D. R., and C.M. Rogers. 1988. Nocturnal energy expenditure of Dark-eyed Juncos roosting in Indiana during winter. *Condor* 90:107–112.

Webster, M. D. 1999. Verdin (*Auriparus flaviceps*). In *The Birds of North America*, no. 470 (A. Poole and F. Gill, eds.) Philadelphia: The Birds of North America, Inc.

Weeks, H. P., Jr. 1994. Pre-laying nest roosting in the Eastern Phoebe: An energy-conserving behavior? *J. Field Ornith.* 65(1):52–57.

Weidensaul, S. 2021. *A World on the Wing: The Global Odyssey of Migratory Birds*. New York: W. W. Norton & Company.

Weimerskirch, H., C. Bishop, T. Jeanniard-du-Dot, A. Prador, and G. Sachs. 2016. Frigate birds track atmospheric conditions over months-long transoceanic flights. *Science* 353(6924):74–78.

Weimerskirch, H., O. Chastel, C. Barbaud, and O. Tostain. 2003. Frigatebirds ride high on thermals. *Nature* 421:333–334.

Wellmann, A. E., and C.T. Downs. 2009. A behavioural study of sleep patterns in the malachite sunbird, Cape white-eye and fan-tailed widowbird. *Animal Behaviour* 77(1):61–66.

White, F. N., G. A. Bartholomew, and T. R. Howell. 1975. The thermal significance of the nest of the Sociable Weaver *Philetairus socius*: Winter observations. *Ibis* 117:171–179.

White, G. 1951. *The Natural History of Selborne*. London: Lutterworth Press.

Whittaker, D. J., and J. C. Hagelin. 2021. Female-based patterns and social function in avian chemical communications. *J. Chemical Ecology* 47:43–62.

Wikelski, M., E. M. Tarlow, C. M. Eising, T.G.G. Groothuis, and E. Gwinner. 2006. Do night-active birds lack daily melatonin rhythms? A case study comparing a diurnal and a nocturnal-foraging gull. *J. Ornithology* 147:107–111.

Wiles, G. J. 1998. Records of communal roosting in Mariana Crows. *Wilson Bull.* 110(1): 126–128.

Williams, J. D., and B. I. Tieleman. 2005. Physiological adaptations in desert birds. *Biosciences* 55:416–425.

Williams, J. G. 1959. Nectarinia johnstoni: A revision of the species, together with data on plumages, moults and habits. *Ibis* 93:579–595.

Williams, T. D. 1995. *The Penguins: Spheniscidae*. Oxford: Oxford U. Press.

Willis, E. O., and Y. Oniki. 2003. Roosting and nesting of the Burnished-buff Tanager (*Tangara cayana*) in southeastern Brazil. *Ornith. Neotropical* 14:279–283.

Woinarski, J.C.Z., B. P. Murphy, S. M. Legge, and ten others. 2017. How many birds are killed by cats in Australia? *Biol. Cons.* 214:76–87.

Wolf, B. O., A. E. McKechnie, C. J. Schmitt, Z. J. Czenze, A. B. Johnson, and C. C. Witt. 2020. Extreme and variable torpor among high-elevation Andean hummingbird species. *Biology Letters* 16:20200428.

Woodall, P. F. 2001. Family Alcedinidae (Kingfishers). In *The Handbook of the Birds of the World*, vol. 6 (J. del Hoyo, A. Elliott, and J. Sargatal, eds.), pp. 130–249. Barcelona: Lynx Edicions.

Woods, C. P., Z. J. Czenze, and R. M. Brigham. 2019. The avian "hibernation" enigma: Thermoregulatory patterns and roost choice of the common poorwill. *Oecologia* 189:47–53.

Wright, J., R. E. Stone, and N. Brown. 2003. Communal roosts as structured information centres in the Raven, *Corvus corax*. *J. Anim. Ecol.* 72(6):1003–1014.

Xu, X., and M. A. Norell. 2004. A new troodontid dinosaur from China with avian-like sleeping posture. *Nature* 431:838–840.

Yamashita, C. 1987. Field observations and comments on the Indigo Macaw (*Anodorhynchus leari*), a highly endangered species from northeastern Brazil. *Wilson Bull.* 99(2):280–282.

Ydenberg, R. D., D. Dekker, G. Kaiser, P.C.F. Shepherd, L. E. Ogden, K. Rickards, and D. B. Lank. 2010. Winter body mass and over-ocean flocking as components of danger management by Pacific dunlins. *BMC Ecology* 10:article number 1.

Ydenberg, R. D., and H.H.T. Prins. 1984. Why do birds roost communally? In *Coastal Waders and Wildfowl in Winter* (P. R. Evans, J. D. Goss-Custard, and W. G. Hale, eds.), pp. 123–139. Cambridge: Cambridge U. Press.

Ydenberg, R. D., H.H.T. Prins, and J. van Dijk. 1983. The post-roost gatherings of wintering Barnacle Geese: Information centres? *Ardea* 71:125–131.

Yom-Tov, Y. 1979. The disadvantage of low positions in colonial roosts: An experiment to test the effect of droppings on plumage quality. *Ibis* 121:331–333.

Zahavi, A. 1971. The function of pre-roost gatherings and communal roosts. *Ibis* 113:106–109.

Zammuto, R. M. and E. C. Franks. 1981. Environmental effects on roosting behavior of Chimney Swifts. *Wilson Bull.* 93(1):77–84.

Zharikov, Y. and D. A. Milton. 2009. Valuing coastal habitats: predicting high-tide roosts of non-breeding migratory shorebirds from landscape composition. *Emu* 109:107–120.

INDEX

Note: Page numbers in *italic* type indicate illustrations.

Emu, 90

energy conservation: balancing of, against energy expenditures, 39, 53, 63, 80, 88; roosting choices for, 76–77, 80, 82; sleep postures for, 28; thermoregulation for, 2, 13, 30, 52–61; weather conditions requiring, 38–39

energy expenditures: balancing of, against energy conservation, 39, 53, 63, 80, 88; balancing of, against predator threats, 71, 74; of migratory birds, 89; of swifts, 126

Eolophus roseicapillus. See Galah

Eremophila alpestris. See Lark, Horned

Erithacus rubecula. See Robin, European

Eudocimus albus. See Ibis, White

Eudocimus ruber. See Ibis, Scarlet

Eudyptula minor. See Penguin, Little

Eumomota superciliosa. See Motmot, Turquoise-browed

Euphagus carolinus. See Blackbird, Rusty

Euphonia, Spot-crowned, 252

Euphonia imitans. See Euphonia, Spot-crowned

Euplectes afer. See Bishop, Golden

Euplectes ardens. See Widow, Red-collared

Euplectes orix. See Bishop, Red

Eupsittula canicularis. See Parakeet, Orange-fronted

Eurocephalus anguitimens. See Shrike, Southern White-crowned

Eurostopodus argus. See Nightjar, Spotted

evolution: of aerial roosting, 122; and sleep posture, 22; sleep's role in, 3, 5, 30

extinction, 29, 184, 266–67

eyes: early feeding in relation to size of, 208; features of, noticeable during sleep, 29–30; magic, 29–30; singing in relation to size of, 198; sleeping with one open, 9–11, 21–22, 25, 30, 122–23

Fairywren, Splendid, 152–53

Falco amurensis. See Falcon, Amur

Falco biarmicus. See Falcon, Lanner

Falco columbarius. See Merlin

Falco cuvierii. See Hobby, African

Falco femoralis. See Falcon, Aplomado

Falcon, Amur: aerial roosting of, 130; human predation on, 267; numbers of, 183, 189

Falcon, Aplomado, 238

Falcon, Barbary, 241

Falcon, Lanner, 241–42

Falcon, Peregrine: flock roosts of, 168; as predator, 71, 169, 173, 220, 238, 240–41

falcons: flock roosts of, 163; numbers of, 183; as predators, 239–42

Falco pelegrinoides. See Falcon, Barbary

Falco peregrinus. See Falcon, Peregrine

Falco rusticolus. See Gyrfalcon

Falco sparverius. See Kestrel, American

Falco tinnunculus. See Kestrel, Common

family roosting, 145

fat reserves, 53–54

feathers: fluffing of, 28–29, 53; tail, 22–23

feeding. *See* foraging

feeding sites: getting information on, 176–77, 230; roosts' distances from, 199–202

feet. *See* legs and feet

ferrets, 245

ffrench, Richard P., 267

Ficedula hypoleuca. See Flycatcher, Pied

Fieldfare, 186

Finch, Gray-hooded, 113

Finch, House: nests built by, 120; pineal gland in, 11

Finch, Zebra: artificial light's effect on, 265; REM sleep of, 6; sleep's benefits for, 3

finches, 3

finches, African, 175

finches, British: resting behaviors of, 44; sleep posture of, 24

finches, European, 91

fish, 72